TRANSLATIONAL MEDICINE: TOOLS AND TECHNIQUES

Edited by

AAMIR SHAHZAD

President, European Society for Translational Medicine (EUSTM),
Vienna, Austria; School of Medicine University of Colorado, Aurora, CO, USA

ELSEVIER

AMSTERDAM • BOSTON • HEIDELBERG • LONDON
NEW YORK • OXFORD • PARIS • SAN DIEGO
SAN FRANCISCO • SINGAPORE • SYDNEY • TOKYO

Academic Press is an imprint of Elsevier

Academic Press is an imprint of Elsevier
125 London Wall, London EC2Y 5AS, UK
525 B Street, Suite 1800, San Diego, CA 92101-4495, USA
225 Wyman Street, Waltham, MA 02451, USA
The Boulevard, Langford Lane, Kidlington, Oxford OX5 1GB, UK

Notices
Knowledge and best practice in this field are constantly changing. As new research and
experience broaden our understanding, changes in research methods, professional practices,
or medical treatment may become necessary.

Practitioners and researchers must always rely on their own experience and knowledge
in evaluating and using any information, methods, compounds, or experiments described
herein. In using such information or methods they should be mindful of their own safety
and the safety of others, including parties for whom they have a professional responsibility.

To the fullest extent of the law, neither the Publisher nor the authors, contributors, or editors,
assume any liability for any injury and/or damage to persons or property as a matter of
products liability, negligence or otherwise, or from any use or operation of any methods,
products, instructions, or ideas contained in the material herein.

ISBN: 978-0-12-803460-6

British Library Cataloguing in Publication Data
A catalogue record for this book is available from the British Library

Library of Congress Cataloging-in-Publication Data
A catalog record for this book is available from the Library of Congress

For information on all Academic Press publications
visit our website at http://store.elsevier.com/

Working together
to grow libraries in
developing countries

www.elsevier.com • www.bookaid.org

Typeset by TNQ Books and Journals
www.tnq.co.in

Printed and bound in the United States of America

Contents

6. Translational Medicine Case Studies and Reports
ALEXANDRE PASSIOUKOV, PIERRE FERRÉ AND LAURENT AUDOLY

7. Translational Approaches in Alzheimer's Disease
STEPHEN WOOD AND GABRIEL VARGAS

List of Contributors

Adriana Amaro Molecular Pathology, IRCCS AOU San Martino–IST Istituto Nazionale per la Ricerca sul Cancro, Genova, Italy

Giovanna Angelini Molecular Pathology, IRCCS AOU San Martino–IST Istituto Nazionale per la Ricerca sul Cancro, Genova, Italy

Laurent Audoly Pierre Fabre Pharmaceuticals, Toulouse, France

Pierre Ferré Pierre Fabre Pharmaceuticals, Toulouse, France

Parviz Ghahramani Chief Executive Officer, Inncelerex, Jersey City, NJ, USA; Affiliate Professor, School of Pharmacy, University of Maryland, Baltimore, USA, Parviz.Ghahramani@inncelerex.com

Dimitris Kalaitzopoulos Oracle UK, Health Sciences Global Business Unit, Reading, UK

Ross D. LeClaire The Translational Bridge, LLC, Albuquerque, NM, USA

Elizabeth K. Leffel Leffel Consulting Group, LLC, Berryville, VA, USA

Alexandre Passioukov Head of Translational Medicine, Pierre Fabre Pharmaceuticals, Toulouse, France

Ketan Patel Oracle UK, Health Sciences Global Business Unit, Reading, UK

Andrea Petretto Core Facility, Istituto G. Gaslini, Genova, Italy

Ulrich Pfeffer Molecular Pathology, IRCCS AOU San Martino–IST Istituto Nazionale per la Ricerca sul Cancro, Genova, Italy

Benedikte Serruys Department of Pharmacodynamics & Translational Medicine, Ablynx, Ghent-Zwijnaarde, Belgium

Thomas Stöhr A2M Pharma, Monheim, Germany

Hans Ulrichts Department of Pharmacodynamics & Translational Medicine, Ablynx, Ghent-Zwijnaarde, Belgium

Maarten Van Roy Department of Pharmacodynamics & Translational Medicine, Ablynx, Ghent-Zwijnaarde, Belgium

Katrien Vanheusden Department of Pharmacodynamics & Translational Medicine, Ablynx, Ghent-Zwijnaarde, Belgium

Gabriel Vargas Neuroscience Early Development, Amgen Inc., Thousand Oaks, CA, USA

Stephen Wood Neuroscience Discovery Research, Amgen Inc., Thousand Oaks, CA, USA

Erfan Younesi Fraunhofer Institute for Algorithms and Scientific Computing, Bioinformatics Department, Schloss Birlinghoven, Sankt Augustin, Germany

About the Editor

Dr Shahzad is currently serving as the president for the European Society for Translational Medicine. Moreover, he is the chairman, Steering Committee for the Global Translational Medicine Consortium. Dr Shahzad is a management committee member of the European Commission's COST action to Focus and Accelerate Cell-based Tolerance-inducing Therapies (A FACTT) and also for the European Commission's COST action on the Development of a European-based Collaborative Network to Accelerate Technological, Clinical and Commercialization Progress in the Area of Medical Microwave Imaging. Dr Shahzad is affiliated with the School of Medicine, University of Colorado, USA. He is visiting professor at the Basic Medical School, Harbin Medical University, and also holds visiting professorship at the First Affiliated Hospital, Harbin Medical University. Dr Shahzad is serving as an editor-in-chief for the "New Horizons in Translational Medicine" (NHTM) and "Translational Medicine Case Reports" (TMCR) journals, published by the Elsevier. Dr Shahzad has advised and participated in establishing translational medicine infrastructure for several organizations. Dr Shahzad has organized several international conferences and is invited chair for numerous international life sciences conferences.

Preface

In recent years, Translational Medicine (TM) has emerged as a powerful interdisciplinary field. To help clarify the many facets of TM the European Society for Translational Medicine (EUSTM) defined TM as an *interdisciplinary branch of the biomedical field supported by three main pillars: benchside, bedside, and community. The goal of TM is to combine disciplines, resources, expertise, and techniques within these pillars to promote enhancements in prevention, diagnosis, and therapies.* Thus, the primary objective of TM is to combine available resources within the individual pillars in order to improve the global health care system.

Translational Medicine: Tools and Techniques is a further initiative of the EUSTM to provide the scientific community with concise knowledge about TM tools and techniques. The initiative was undertaken to reduce confusion about techniques, tools, and applications. This book is intended to help professionals both in academia and industry as well as students and young investigators perusing careers in TM field.

The book is divided into seven chapters and written by the internationally respected authors from both academia and industry. New approaches for biomarkers discovery, development, and validations are discussed in the Chapter 1. Chapter 2 presents advancements in data mining and management tools. Chapter 3 discusses the modeling and simulation applications in drug development process. Advancements in omics sciences are described in Chapter 4. Chapter 5 provides an overview of the regulatory process in United States of America, Europe, China, and Japan. A pearl of this book is the inclusion of case reports and studies, which will help the reader better understand TM applications. Chapters 6 and 7 include translational medicine case studies and reports.

I am thankful to all the authors for their valuable time and contributions for this timely book. Moreover, I am very grateful to Ms Mica Haley and Ms Lisa Eppich from Elsevier for their continuous support and kindness during all stages of book production; without their help the publication in such a short duration of time would not have been possible.

Aamir Shahzad
October 2015

Acknowledgments

Prof. Randall J. Cohrs
USA

Prof. Gottfried Köhler
Austria

This book is dedicated to Prof. Randall J. Cohrs and Prof. Gottfried Köhler. Randall's application of basic molecular virology findings to clinical problems demonstrates the many facets within translational medicine. Gottfried's scientific journey from biophysics to molecular diagnostics is an inspiration to all endeavoring to succeed in translational medicine. Together, their excellent contributions and achievements in their fields, continuous encouragement, and support are always a source of inspiration. Special thanks to Ms Sandra Oberhuber for her wonderful support. The Acknowledgments section would remain incomplete without mentioning my family: Mahrose Aamir and Sarah Shahzad as they sacrifice their time for completing the book.

1

New Developments
in the Use of Biomarkers
in Translational Medicine

*Benedikte Serruys[1], Thomas Stöhr[2], Hans Ulrichts[1],
Katrien Vanheusden[1], Maarten Van Roy[1]*

[1]Department of Pharmacodynamics & Translational Medicine, Ablynx,
Ghent-Zwijnaarde, Belgium; [2]A2M Pharma, Monheim, Germany

O U T L I N E

Translational Medicine: Tools and Techniques
http://dx.doi.org/10.1016/B978-0-12-803460-6.00001-5

1

INTRODUCTION

A biomarker can be defined as a characteristic that is objectively measured and evaluated as an indicator of normal biological processes, pathogenic processes, or pharmacologic responses to a therapeutic intervention [1]. The first publication associated in PubMed with the search term "biomarker" dates back almost 70 years [2]. Since then there has been an explosion of published research on biomarkers yet with limited translation into clinical practice and back to bench.

Classically, biomarker experiments involved the assessment of one or several proteins in the blood of a group of patients or experimental animals versus a control group. In recent years there have, moreover, been many developments in biomarker research that makes biomarkers an extremely valuable tool in translational medicine.

This chapter is not intended to provide a general review on the use of biomarkers in translational medicine. For this the reader is referred to a number of recent review articles or books [96–98]. The goal of this chapter is to highlight a number of recent developments that advanced the field and may help to value the sophisticated use of biomarkers in biomedical research.

The selection of biomarkers should normally be part of a translational strategy (Section Biomarkers as Part of a Translational Strategy). Rather than measuring proteins in blood, biomarker assessment can be done locally in the target tissue or even intracellularly (Section From Blood- to Tissue-Specific Biomarkers). Biomarker measurements can include functional assays (Section From Static to Functional Biomarker Assays), in vivo imaging (Section From Static Ex Vivo Monitoring to In Vivo Continuous Imaging), or even abstract measures such as a quality of life score (Section From Peptide Molecules to Nonbiochemical Biomarkers). Moreover, the analysis of biomarker results can be done using algorithms (Section From Single Biomarkers to Biomarker Patterns) or with modeling (Section From Describing the Status Quo to Predicting/

Modeling the (yet) Unknown). To be of real value, biomarker results need to be sufficiently documented to enable further clinical development (Section From Isolated Reporting to Meaningful Representation of Results that Allows Replication and Confirmation). The usefulness of biomarkers has been clearly demonstrated in the area of patient stratification, especially in the field of oncology, and this is slowly progressing to get established in more heterogeneous autoimmune diseases as well (Section From the Discovery of Pathomechanisms to the Definition of Patient (Sub)populations).

In the next sections (overview see Table 1), we will highlight these developments and give particular emphasis to recent examples from the literature.

TABLE 1 Chapter Overview

Area of recent biomarker developments		Section
Biomarker selection	Embedding in a translational strategy	Biomarkers as Part of a Translational Strategy
New types of biomarkers	Target tissue and intracellular biomarkers	From Blood- to Tissue-Specific Biomarkers
	Functional biomarker assays	From Static to Functional Biomarker Assays
	In vivo imaging	From Static Ex Vivo Monitoring to In Vivo Continuous Imaging
	From peptide molecules to nonbiochemical biomarkers	From Peptide Molecules to Nonbiochemical Biomarkers
New ways of analyzing biomarker data	Biomarker patterns	From Single Biomarkers to Biomarker Patterns
	Biomarker modeling	From Describing the Status Quo to Predicting/Modeling the (yet) Unknown
Biomarker documentation	Reporting standards	From Isolated Reporting to Meaningful Representation of Results that Allows Replication and Confirmation
Promising applications	Patient stratification	From the Discovery of Pathomechanisms to the Definition of Patient (Sub) populations

BIOMARKERS AS PART OF A
TRANSLATIONAL STRATEGY

Biomarkers are valuable tools when used fit for purpose to answer specific questions during the course of clinical drug development. For instance, a particular biomarker may be suitable to answer one specific question (e.g., Does the drug candidate show direct target engagement in a certain study?) but may not be useful to answer another question (e.g., Does the drug candidate show efficacy in that same or in another study?). An answer to specific questions during translational research is often generated via an isolated approach, i.e., by focusing on one single experiment and questioning which biomarker is the best candidate for this specific experiment. A higher translational value would be obtained by embedding the biomarker question into a more long-term translational strategy, by investigating the question in multiple studies, going from in vitro experiments over ex vivo studies to in vivo preclinical studies to finally clinical trials. The ultimate goal of this integrated approach is to increase (pre)clinical development efficiency and to decrease the failure risk in the process of bringing products to the market. The main prerequisites for a successful translational strategy are (1) a high translational value of the in vitro/ex vivo/in vivo models and (2) a high translational value of the specific biomarkers. However, translational challenges in biomarker measurements make it difficult to bridge results from the preclinical to clinical setting (and back). For example, efficacy (of intervention) biomarkers are generally more difficult to translate between ex vivo/in vivo models and the human situation as these correlate with a favorable outcome by drug treatment in the specific species and the model (i.e., efficacy) and are therefore often less closely linked to the direct action of the drug in man. To illustrate, imaging techniques in osteoarthritis (OA) focus on in-life assessment in the clinic, in contrast to terminal assessment via histopathology in animals. Also, pain assessment is difficult to translate, as pain is mainly monitored by self-reporting questionnaires in clinical research, but these cannot be readily mirrored in animal studies. In contrast, pharmacodynamic (PD) biomarkers are often more translational, as they reflect direct target engagement.

Overall, the ideal biomarker strategy starts with considering the target product profile of the compound and translating this backward. If, for instance, a superior efficacy is the ultimate goal, then the translational strategy should determine how this efficacy can be best measured with the help of biomarkers and in which particular step of the development. This way of thinking with the long-term goal in mind may help to select more relevant biomarkers and to enhance their translational value.

To answer a specific question by biomarkers without a lot of up front information on which biomarkers are appropriate for this objective, it is recommended to start preclinical development with a range of candidate biomarkers, which are being investigated in parallel in pilot in vitro/

ex vivo/in vivo studies. Based on this early investment, the most promising biomarkers are selected to investigate further in more pivotal preclinical studies and later on in clinical trials. Along the way, a biomarker can serve to answer multiple questions. For example, a PD biomarker can turn out to be a relevant stratification biomarker, which may lead to the development of a companion diagnostic.

The clinical development plan benefits from a clear translational biomarker strategy: first, to design the first-in-man studies, and second, to translate the data from the clinical trials back to the bench in order to better understand the clinical data. In turn, the newly generated bench data can serve the more advanced clinical trials or can guide the development of follow-up molecules. The latter can be nicely illustrated by the development path of the anti-CD40L monoclonal antibody (mAb) ruplizumab and its follow-up molecule CDP7657 (see Figure 1). Their target CD40L plays a role in the activation of professional antigen-presenting cells and is expressed on activated T cells [3]. During clinical trials, however, thromboembolic (TE) events were observed. In a lupus nephritis trial, two patients suffered a myocardial infarction [4]. Due to these observations, the study was prematurely terminated. In a transplant trial, one obese, bedridden woman died from pulmonary embolism [5]. Around that time, a Crohn disease patient developed a blood clot in her leg during a Phase II study receiving another

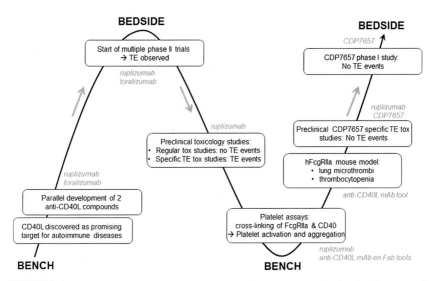

FIGURE 1 **Illustration of a translational biomarker strategy with the development of ruplizumab/toralizumab and the follow-up molecule CDP7657.** Information from the clinical trials ("bedside") of ruplizumab and toralizumab was used to feed an extensive preclinical investigation ("bench"), rendering new insights in the CD40L target biology and mode of action of the candidate therapeutic compounds. Based on this new information, a follow-up molecule CDP7657 is currently under development. During its preclinical and early clinical development, special attention has been paid to platelet and thromboembolic (TE)-related biomarkers.

CD40L-targeting compound toralizumab [5]. Initially, these observations could not readily be explained. In theory, the drug could also bind and activate platelets as CD40L is also expressed on this cell type, but there was at the time no evidence for this [5]. By going back "from bedside to bench," evidence has been generated showing that these TE events are most likely caused by a dual interaction between the platelets and the anti-CD40L antibody. On the one hand, there is indication of cross-linking by human FcgRIIa, i.e., subtype of Fc receptors, on platelets and the Fc part of the anti-CD40L antibody. On the other hand, cross-linking likely also occurs by CD40 on platelets and soluble CD40L, which is bound by the anti-CD40L antibody. This dual cross-linking is shown to result in ex vivo platelet activation and ex vivo platelet aggregation [6–9]. This ex vivo information was used to set up a transgenic mouse model, expressing human FcgRIIa since mice do not express this receptor. Addition of preformed complexes of recombinant soluble CD40L and anti-CD40L antibody to these mice generated thrombocytopenia and intravascular microthrombi [7]. The hypothesis seemed to be further confirmed by the data obtained with ruplizumab and the follow-up molecule CDP7657 in rhesus monkeys [10]. CDP7657 is a PEGylated anti-CD40L Fab fragment, assuming that the lack of a functional Fc domain would avoid the development of TE events. A lower prevalence of microthrombi in the lungs of rhesus monkeys was indeed demonstrated following treatment with CDP7657 as compared to ruplizumab. The number of microthrombi in animals treated with CDP7657 was similar to the number in vehicle-treated animals and historical controls. The same holds true for variants of ruplizumab without effector function such as the aglycosylated version of the mAb and its di-Fab fragment. Moreover, no TE events were reported in a Phase I study with CDP7657 in both healthy individuals and patients with systemic lupus erythematosus (SLE) [11].

Taken together, CDP7657 development is a good illustration of how the development of clinical compounds may benefit from an integrated translational biomarker approach. Indeed, the bedside-to-bench translation based on ruplizumab data has provided information to guide the development of CDP7657.

NEW TYPES OF BIOMARKERS

From Blood- to Tissue-Specific Biomarkers

Blood is a central compartment in which many signaling molecules are secreted, often after having served their primary signaling purpose locally. Detection of biomarkers in blood has the clear advantage of avoiding invasive procedures but can lack the required specificity and sensitivity since the origin of the measured signaling molecules cannot be identified and the concentration is more diluted than locally. Measuring biomarkers locally, in the specific tissues, might therefore have a specific advantage.

Many examples of tissue-specific biomarkers are found in the field of oncology. In particular, use of immunohistochemical biomarker assessment in solid tumors has rapidly progressed alongside the development of the tissue microarray and imaging technologies. Moreover, automated image analysis systems with more precise quantitation are now progressively replacing the subjective, semiquantitative manual scoring previously performed by pathologists.

An example of the use of blood- and tissue-specific blood markers can be found in the field of prostate cancer [12,13]. This disease remains the second leading cause of cancer-relating deaths in men. In clinical practice, measurement of serum total prostate-specific antigen (PSA) levels has been the "gold" standard in determining the presence and stage of prostate cancer. PSA is a protein produced only by prostate tissue and therefore measurement of serum levels of this marker would seem to be specific for prostatic disease. PSA is used as a diagnostic marker for both early detection of prostate cancer and for follow-up after surgery or during treatment. The US Food and Drug Administration (FDA) officially approved the use of PSA for prostate cancer screening in 1994 and defined 4.0 ng/mL as the upper limit of normal. However, at the time, no randomized controlled trial of prostate cancer screening was performed, with the main risk of overdetection [14], bringing subjects with indolent disease at risk upon surgery. The value of PSA as a widespread screening tool was later assessed by two large population-based trials. In the first trial, "Prostate, Lung, Colorectal, and Ovarian (PLCO) Cancer Screening Trial" [15], no effect on prostate cancer mortality through yearly PSA testing and digital rectal examination could be observed. The European Randomized Study of Screening for Prostate Cancer, on the other hand, did demonstrate that a PSA-based screening could lead to a 20% relative reduction of prostate cancer-induced death although with a high incidence of overdiagnosis [16], which was confirmed after further follow-up [17]. Based on these results, the US Preventive Services Task Force recommended against PSA-based screening for prostate cancer [18].

As a consequence, the quest for better prognostic biomarkers in prostate cancer is still very active. Specific proteins in tissue specimens taken during diagnostic biopsy or radical prostatectomy could become useful for diagnosis and prognosis but could also be interesting for the development of targeted therapies. One example of these markers is prostate stem cell antigen (PSCA), a cell surface protein of normal prostate tissue, which is highly expressed in primary prostate tumors and related bone metastases [19] but with limited expression in extraprostatic normal tissues [20]. Expression of PSCA is measured through histological staining for the protein or through quantification of its mRNA. Expression levels increased with other tumor characteristics such as higher Gleason score, tumor stage, and progression to androgen independence. PSCA has been shown to be a predictor of cancer progression in patients with benign prostatic hyperplasia after transurethral resection [21]. Moreover, based on these investigations, PSCA is also under investigation for targeted therapies. Indeed, PSCA was used as an antigen for dendritic cell-based immunotherapy in hormone- and chemotherapy-refractory prostate cancer [22].

As a consequence of the evolution in tissue-based biomarker research, automated image analysis is replacing manual image analysis in order to improve accurate quantification and standardization. One example of such automated image analysis is the establishment of an Immunoscore in oncology as a clinically useful prognostic marker [23]. This Immunoscore is based on the enumeration of two lymphocyte populations, (CD3/CD45RO, CD3/CD8, or CD8/CD45RO), both in the center and in the invasive margin (IM) of tumors. Increasing literature has demonstrated that the progression of cancer is influenced by the host immune system and that tumors are often infiltrated by various numbers of lymphocytes, macrophages, or mast cells. A histological scoring of these features might be superior as a prognostic marker, in particular compared to the traditional classification tools, which are solely dependent on the analysis of tumor cells. The Immunoscore ("I") relies on the identification of the center and the IM of the resected tumors and the subsequent enumeration of CD8 and CD45RO cells in these regions to provide a score ranging from Immunoscore 0 ("I"0), when low densities of both cell types are found in both regions, to Immunoscore 4 ("I"4), when high densities are found in both regions (Figure 2). When applied to two large independent cohorts (n=602), the Immunoscore revealed a high prognostic value: only 4.8% of patients with a high "I"4 relapsed after 5 years and 86.2% were still alive.

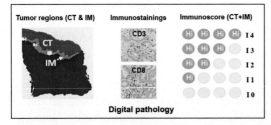
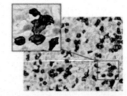

FIGURE 2   **Calculation of Immunoscore.** The Immunoscore consists of a precise quantification of the combination of two lymphocyte markers (CD3+ and CD8+) in two tumor regions (center of tumor (CT) and invasive margin (IM)). An automated detection of the tumor IM by digital image analysis is performed after manual annotation of tumor regions. Precise quantification is performed on whole slide sections using digital pathology and specific algorithms integrating, shape, morphology, staining intensity, cell location, and counts. The Immunoscore is currently being validated in a worldwide clinical study. *From Ref. [23].*

In comparison, 72% of patients with a low score ("I"0 and "I"1) experience tumor recurrence and only 27.5% were still alive at 5 years [24].

From Static to Functional Biomarker Assays

Biomarkers that reflect the (patho)physiological status of a system, or a pharmacological effect on that system, can be of more translational value than the quantification of circulating levels of protein markers, or gene expression. Static biomarker assays (such as the quantification of a cyto-kine in plasma levels) would rather reflect a snapshot of a certain (patho) physiological situation in a biological system, while functional biomark-ers (such as ex vivo cytokine release assays) can reflect the possibility of a biological system to react to stimuli. However, the related functional biomarker assays are often more challenging to develop or validate and can require specific equipment. Moreover, consideration of preanalytical variables and method standardization is of key importance for assuring a correct analysis within or across studies.

The first example of functional biomarker assays can be found in the field of thrombosis and hemostasis: platelet function tests are used for the pre-diction of hemorrhage in presurgical/perioperative settings, or for the monitoring of antiplatelet therapy to identify hypo- or hyperresponders at respective risk for thrombosis or hemorrhage [25]. These functional tests are considered as superior in terms of sensitivity and specificity compared to static tests, which measure, for instance, soluble markers of activated platelets (e.g., soluble P-selectin). A wide variety of platelet function tests have been developed to address different functions of platelets, includ-ing light transmission platelet aggregometry, which was developed in the 1960s [26] and is still considered as the "gold" standard. This methodology assesses platelet-to-platelet clump formation induced by an exogenous platelet agonist. The clumping is quantified as increase of light transmis-sion through an optically dense sample of platelet-rich plasma. Although this methodology is the most widely employed method to monitor anti-platelet therapies and accepted as the most complete assay in terms of platelet functionality, it needs to be performed by a laboratory staff who has a high degree of skill, experience, and expertise. As a consequence, various attempts have been made in standardizing this methodology. In particular the procedures to prepare the platelet-rich plasma sample are considered an essential preanalytical variable as incorrect sample process-ing might affect platelet function and—as a consequence—might con-found the test results [27]. As light transmission platelet aggregometry is hampered for routine use because of the reasons mentioned above, newer devices were developed that allow a more standardized analysis of plate-let aggregation. The multiple electrode aggregometry (MEA) methodol-ogy, based on the multiple platelet function analyzer device, is a test that measures platelet function by using citrated blood as a matrix without sample processing [28]. Platelet aggregation is measured as a change in impedance, generated by an agonist-induced adhesion and clumping of platelets to artificial surfaces of two electrodes. The MEA technique has been used to identify nonresponders to antiplatelet therapy or patients who are at risk of bleeding under such therapy. A prospective study in 1068 patients demonstrated the predictive value of MEA for stent thrombosis [29]. Subjects undergoing percutaneous coronary intervention were given the antiplatelet drug clopidogrel before the procedure. Low responsiveness to clopidogrel assessed through MEA was linked with a higher risk for the occurrence of stent thrombosis. Similarly, a retrospective study in patients under thienopyridine therapy demonstrated that the MEA technique was predictive of postoperative clinical bleeding [30].

A second example of functional biomarker analysis is intracellular cytokine staining (ICS). This is a flow cytometry-based technique, which consists of the simultaneous detection of the phenotype of an individual cell and the characterization of its cytokine production after short-term stimulation [31]. It is the most commonly used technique to measure T-cell reactivity and is frequently employed as a measure of immunogenicity as efficacy end point in vaccination trials. The method is typically performed on whole blood or peripheral blood mononuclear cells (PBMC) with no prior enrichment of cell populations. The method has been standardized and can be validated [32]. Similarly as for the platelet function tests, standardization of the procedures used for sample processing is of utmost importance to compare efficacy of vaccine candidates between multiple clinical trials. Specifically, length of time from venipuncture to cryopreservation (in the case of PBMC) has been identified as the most important parameter influencing T-cell performance in the ICS methodology [33]. ICS has been used in various clinical studies, in particular in those assessing the CD4+ and CD8+ T-cell responses of various HIV vaccine candidates [34–36]. However, the correlation of the functional response in the ICS assay with HIV vaccine efficacy is still under debate. This concern was highlighted within the Step study, which assessed the efficacy of a cell-mediated immunity vaccine against HIV-1. This study was terminated early due to lack of efficacy and increased incidence of HIV-1 infection in vaccine-treated compared to placebo-treated men with preexisting adenovirus serotype 5 immunity [37], despite the high-vaccine immune response measured through ICS [38].

From Static Ex Vivo Monitoring to In Vivo Continuous Imaging

In Section From Blood- to Tissue-Specific Biomarkers, the advantages of target tissue-based biomarkers have been highlighted. One of the inherent challenges of this kind of biomarkers is the sampling of the tissue, which is, depending on the tissue, more or less invasive. In addition, sampling in the same tissue is restricted to one or at maximum a few samples, thus the temporal resolution is less than optimal. One of the organs least accessible for direct tissue sampling is the brain.

Neuroimaging techniques like magnetic resonance imaging (MRI) and positron emission tomography (PET) imaging have proven, over the last two decades, to be useful tools to enhance the accuracy of clinical diagnosis in neurodegenerative diseases and to support disease progression monitoring. Noninvasive PET imaging has greatly contributed to a better understanding of neurologic diseases like Alzheimer disease (AD) and Parkinson disease and will support the clinical development of future disease-modifying drugs for these and other indications.

AD is characterized by plaques consisting of beta amyloid protein, neurofibrillary tangles (NFT) of phosphorylated tau protein, neurodegeneration, and ultimately the clinical symptoms of cognitive impairment and dementia. Currently, no true disease-modifying drugs exist and existing therapies only temporarily halt or improve the process of cognitive decline in these patients [39].

In vivo PET imaging is the most accurate and sensitive technique to evaluate the neurodegenerative processes early on in the patients, even years before clinical symptoms are manifested. PET tracers identify functional changes in brain metabolism and protein level alterations, e.g., tracers for amyloid protein, tau, and glucose metabolism have provided new insights that challenge the amyloid hypothesis, the predominant theory for the cause of AD, which stipulates that the deposition of beta amyloid is the initial event in AD pathogenesis. PET imaging has provided the evidence that the initial appearance of brain injury is not per definition linked to presence of beta amyloidosis and conversely, that beta amyloid load does not warrant clinical signs of dementia, especially in older people [40]. In addition, beta amyloid plaques and NFT can develop independently and in random order [41].

Another imaging marker has proven more promising for early detection and monitoring of cerebral dementia. Fluorodeoxyglucose (FDG)-PET visualizes cerebral metabolism as it is a glucose analog and reveals in AD a lower neuronal energy demand in certain areas (as represented in Figure 3). This is interpreted as a measure of impairment of neuronal function (i.e., hypometabolism) [42] and has been shown to correlate with cognitive impairment. As a change in cerebral metabolism is an early marker in the cascade leading to dementia and has been reported in patients at risk, it is an interesting biomarker to support early diagnosis of AD patients and to monitor the evolution of the disease. Amyloid-PET has been performed in thousands of AD patients with either 18F-labeled tracers (e.g., florbetapir and florbetaben) or Pittsburgh Compound-B (PiB; see Figure 3). These amyloid beta tracers are specific and have proven

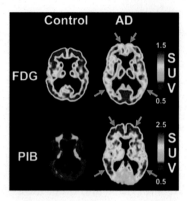

FIGURE 3   In vivo imaging of amyloid and hypometabolism: representative PiB and FDG scans from control and AD participants. Arrows indicate areas of typical hypometabolism (FDG) or typical amyloid deposition [47].

TABLE 2   Equations and Definitions for Sensitivity, Specificity, Positive Predictive Value, and Negative Predictive Value

		Truth	
		Disease	**Nondisease**
Test result	Positive	**A** (true positive)	**B** (false positive)
	Negative	**C** (false negative)	**D** (true negative)
Sensitivity	$A/(A+C) \times 100$	Sensitivity is the probability for a positive test result for those with the disease	
Specificity	$D/(D+B) \times 100$	Specificity is the probability for a negative test result for those without the disease	
Positive predictive value (PPV)	$A/(A+B) \times 100$	The proportion of patients correctly diagnosed with the disease	
Negative predictive value (NPP)	$D/(D+C) \times 100$	The proportion of patients correctly diagnosed not having the disease	

utility in mapping the beta amyloid load in multiple Phase III clinical trials [43,44]. Although amyloid beta protein has long been the hallmark of AD and is considered a good diagnostic marker, it is unsuitable for follow-up of disease progression as in later stages of the disease it plateaus when neurodegeneration and atrophy further increase together with cognitive decline. In addition, in elderly patients the positive predictive value (PPV; see Table 2) of amyloid-PET is under discussion [45]. The Tau-PET tracers have endured more selectivity issues than the beta amyloid-specific tracers but are of great value to extend the understanding of the different processes that contribute to AD. Hyperphosphorylation of tau results in the formation of NFT and is inherent to multiple neurodegenerative diseases. The tau tangle growth, however, reflects well the evolution of the clinical symptoms; this in contrast to the beta amyloid load [46].

Key players in the pathology of AD have been identified in beta amyloid, phosphorylated tau, and glucose metabolism, highlighting that the pathology of AD cannot be adequately described by one biomarker and this will be the case for most neurological diseases. The AD neuroimaging initiative has been validating biomarkers for clinical trials and diagnosis of AD and has made all this information available to help progress the treatment of AD patients and encourage new investigations.

In brief, PET imaging has provided crucial tools for in vivo monitoring and describing the pathology of neurological diseases. It is believed that PET imaging will have a significant impact on future clinical trials by (1) facilitating better characterization of the patients as imaging might enable a more homologous patient population leading to higher chance of reaching statistical significance; (2) choosing the right patients for a particular compound or selecting patients in a particular phase of the disease; (3) making apparent whether a drug is hitting its target and is worthwhile developing; (4) helping with faster and more accurate diagnosis of patients with a neurological disease, ideally identifying patients before irreversible damage has occurred; and (5) functioning as outcome measurements for trials, complementary to current clinical outcome measures.

From Single Protein Molecules to Nonbiochemical Biomarkers

Biomarkers are often viewed in a narrow sense, i.e., as specific biochemical molecules being proteins or peptides which are measured in body fluids. However, the definition of a biomarker (see Introduction) is much broader and should encompass "any characteristic that is reflecting a biological state and that can be measured objectively." In the field of musculoskeletal disorders such as rheumatoid arthritis (RA) and OA, several types of these nonbiochemical biomarkers are being used. In the next paragraphs, two examples of such biomarkers are described for OA: (1) imaging techniques and (2) quality of life measurements.

The first example of nonbiochemical biomarkers in OA focuses on imaging techniques. These go beyond the PET imaging of specific markers as described in the previous section and offer great potential for the diagnosis of OA, for the assessment of disease severity, for monitoring the disease progression, and for follow-up of treatment efficacy. In this section, the present and future use of imaging techniques as biomarkers for OA is discussed. Imaging techniques as biomarkers illustrate the concept of a complex biomarker, a composite biomarker, which does not rely on the measurement of just one but several parameters.

Next to symptom-based diagnosis of OA, conventional radiography (CR) has been an important diagnostic method, as it is the only imaging technique which is currently clinically accepted for the diagnosis of OA and the monitoring of disease severity and disease progression [48,49]. Several OA-specific bone structures are visualized by CR like osteophytes, subchondral sclerosis, and subchondral cysts. No soft tissues can be visualized, rendering only limited sensitivity for OA pathological changes. Consequently, the method is not useful to detect structural changes in the early stages of OA or to detect small deterioration changes in the OA disease progression process. Also, it implies that radiography is no ideal tool for the early evaluation of disease-modifying treatment candidates [50].

Because of the low sensitivity and discordance with clinical symptoms by CR, major efforts are ongoing to validate MRI as alternative OA imaging technique. In contrast to CR, MRI can visualize and discriminate next to bone, soft tissues like cartilage, menisci, and ligaments, allowing whole-organ imaging of the joint [48,51]. Because of this feature, MRI looks promising to enable earlier OA diagnosis (i.e., in the preradiographic phase), to detect premature changes in the OA progression process, and to allow more quick evaluation of treatment effectiveness. Moreover, MRI shows a much stronger correlation with the incidence and severity of pain than CR [52].

In light of biomarker standardization, the imaging biomarker techniques should be quantitative and objective. Grading scales in OA imaging techniques have been developed to meet these needs. Commonly used grading systems in OA are the Kellgren–Lawrence (KL) grading system and OARSI score system for CR [48,53,54], and the Whole-Organ MRI Score and Boston Leeds Osteoarthritis Knee Score for MRI [55,56]. These scoring systems nicely illustrate the concept of a composite score of several parameters, so not just focusing on one biomarker. For instance, the MRI scoring systems score several parameters, including bone marrow lesion size, synovitis, effusion, meniscus integrity, ligament integrity, cartilage integrity, subarticular bone morphology (including osteophytes, subchondral cysts), and subarticular bone attrition, which ultimately result in one overall composite score.

In conclusion, although still not fully clinically validated, MRI imaging offers great potential as biomarker technique in OA over CR for (1) the diagnosis of the disease, (2) the monitoring of the disease severity and progression, and (3) follow-up of effectiveness of treatment candidates.

The second example of nonbiochemical biomarkers in OA is the quality of life measurements. Pain is one of the most prominent symptoms of patients with musculoskeletal disorders such as RA or OA. Pain has an impact on sleep and physical function (and vice versa) [57]. Thus, pain in particular and quality of life more generally are probably the most significant clinical outcomes and the ultimate goal of every therapy.

Measurement of pain, sleep, and physical function mainly relies nowadays on self-reporting questionnaires. For instance, the Western Ontario and McMaster Universities (WOMAC) Osteoarthritis Index is the most commonly used questionnaire to assess pain, sleep, and activity in patients with knee/hip OA. These self-reporting questionnaires are per definition subjective and present consequently a high intra- and interperson variability. Due to this inherent subjectivity, it can be questioned if self-reporting questionnaires should be seen *sensu strictu* as real biomarkers. More objective measurements for pain, sleep, and physical function are under evaluation. These alternatives should not be viewed as substitute for the questionnaires but as valuable complementary information.

For instance, functional MRI (fMRI) is currently being evaluated to measure pain in OA [58], representing a more objective assessment compared to the pain-related questions in the abovementioned questionnaires. One recent example is a small study with painful hand OA patients receiving naproxen, showing the potential of fMRI to detect changes in brain functioning following treatment with the compound [59].

Complementary to the physical function-related questions in the questionnaires is the inclusion of performance-based outcome measures in OA clinical trials. These tests measure what a person can do and not what a person perceives he can do [60] and is therefore more objective as measurement of activity. Unfortunately, there are a wide variety of performance-based tests and no gold standard test is available yet. Due to the absence of a consensus, the Osteoarthritis Research Society International (OARSI) published in 2013 a recommended set of performance-based measurements to be used complementary to self-reporting questionnaires and which is intended for patients across the entire spectrum of OA severity as well as following hip or knee arthroplasty [61].

In conclusion, self-reporting questionnaires are essential nonbiochemical biomarkers in the OA field, but due to the subjective nature, these are becoming complemented by more objective measurements like fMRI and performance-based measurements.

NEW WAYS OF ANALYZING BIOMARKER DATA

From Single Biomarkers to Biomarker Patterns

A strong opinion in the biomarker field is that no single biomarker can sufficiently capture the complex biology of human diseases like cancer and autoimmune diseases, and that the use of a panel of markers, combined in some type of algorithm, would improve disease diagnosis, tracking of disease activity (DA), and treatment evaluation. Multiplexed immunoassays that provide multiple, parallel protein measurements on the same small biological sample have therefore become popular tools in biomarker discovery research. They have demonstrated that describing or following the pattern of multiple biomarkers has the potential to provide a more sensitive and more specific signal than a single biomarker. The possible advantage of biomarker patterning is further highlighted by two examples: (1) the use of multiplexing for diagnostic purposes in ovarian cancer and (2) a new biomarker algorithm for DA assessment and treatment evaluation in RA.

Ovarian cancer presents with very few, if any, specific symptoms and its associated high mortality rate is due to the lack of an adequate screening strategy to detect early-stage disease. The currently used protocol for early detection of ovarian cancer in the high-risk population is transvaginal ultrasound in combination with screening for elevated levels of the biomarker carbohydrate antigen 125 (CA125) [62,63]. However, this protocol does not reduce mortality compared with usual care, is not cost-effective, and surgery following a false-positive screening test is sometimes associated with complications [62,63]. Although CA125 remains the most useful individual indicator of ovarian cancer, CA125 on its own exhibits only a sensitivity (see Table 2 for definition) of less than 60% in early stages of the disease [63,64]. The search for additional biomarkers that are capable of complementing the performance of CA125 remains therefore important in order to achieve better levels of sensitivity and specificity. A frequently discussed study by Visintin et al. [64] demonstrated that a newly developed algorithm for six protein serum markers (Table 3), measured using a multiplex platform, appears to have a significant better diagnostic reliability compared to serum CA125 alone in discriminating healthy individuals from ovarian carcinoma patients [62,64]. The six biomarkers are all produced either by the surrounding supportive cells or as a response to signals originating from the ovary [62,64]. This panel was made commercially available under the trade name Ova-Sure (LabCorp) with a reported sensitivity and specificity (see Table 2 for definition) of 95.3% and 99.4%, respectively [64], which is much higher compared to the 60% sensitivity reported for CA125 on its own. However, these results have been criticized by other scientists as it was found that the PPV of the test (see Table 2 for definition) was inaccurately calculated [65]. This led to the eventual withdrawal of the kit and also illustrates the challenges facing biomarker development in general. After this setback, research continued and two newly developed algorithms can hopefully generate more reliable screening tools. The first scoring model is the Risk of Ovarian Malignancy Algorithm (ROMA, Fujirebio Diagnostics) (Table 3), which combines the CA125 and human epididymal secretory protein 4 (HE4) levels in serum with the menopausal status of women presenting with a pelvic mass to generate a single numerical score that correlates with the likelihood of malignancy [63]. This model originated from a study in

TABLE 3 Overview of the Here-Discussed Multiple Biomarker Tests in Ovarian Cancer and Rheumatoid Arthritis

BM test	Company	Original BM	Additional BMs	Algorithm	Analytical platform	Regulatory status
OVARIAN CANCER						
OvaSure	LabCorp	CA125	Leptin, prolactin, OPN, IGF-II, MIF	Yes [64]	Multiplex	Withdrawn
ROMA	Fujirebio Diagnostics	CA125	HE4	Yes [72]	CMIA + EIA	Marketed, FDA cleared
OVA1	Vermillion	CA125	ApoA1, TF, B2M, prealbumin	Yes (OvaCalc®)	5 immunoassays	Marketed, FDA cleared
RHEUMATOID ARTHRITIS						
Vectra DA	Crescendo Bioscience	CRP	SAA, IL-6, TNF-RI, VEGF-A, MMP-1, YKL–40, MMP-3, EGF, VCAM-1, leptin, resistin	Yes [68,70]	Multiplex	Commercialized

ApoA1, apolipoprotein 1A; B2M, beta2-microglobulin; CMIA, chemiluminescent microparticle immunoassay; EGF, epidermal growth factor; EIA, enzyme immunometric assay; IGF-II, insulin-like growth factor II; IL-6, interleukin-6; MIF, macrophage inhibitory factor; MMP-1, matrix metalloproteinase 1; OPN, osteopontin; SAA, serum amyloid A; TF, transferrin; TNF-RI, tumor necrosis factor receptor type-1; VCAM-1, vascular cell adhesion molecule 1; VEGF-A, vascular endothelial growth factor A; YKL–40, cartilage glycoprotein 39.

which multiple serum biomarkers in 233 women with a pelvic mass were measured, the results showing that HE4 had the highest sensitivity for detecting ovarian carcinoma, especially stage I disease, and that combining CA125 and HE4 was a more accurate predictor of malignancy than either alone [66]. ROMA was approved by the FDA in 2011, although recent evaluations of ROMA have produced mixed results. A number of studies have reported results which reaffirm the complementary performance of HE4 and CA125 and the superior diagnostic abilities of ROMA over CA125 alone in various patient cohorts, while other groups have reported contrary results showing that variability in the composition of the target population has an impact on the performance of the CA125/HE4 combination [63]. The second biomarker-based diagnostic test is available under the trade name OVA1 (Vermillion) and uses five biomarkers which were identified through serum proteomics using SELDI-TOF-MS. The test was approved by the FDA in 2009 for use as an adjunct to physical examination and imaging and produces a risk assessment score within the range of 0–10 [63,67] (Table 3). Success going forward with these two new algorithms will be measured by the ability of these kits to produce a meaningful reduction in mortality, morbidity, and cost resulting from unnecessary surgical procedures in ovarian cancer.

RA is a chronic inflammatory disorder that primarily affects the joints. Accurate and frequent assessment of RA DA is important for patient management, and current DA measures are typically composite scores that include physician assessment of symptoms, patient-reported outcomes, and laboratory measurements. The Disease Activity Score based on 28 joints (DAS28) is for the moment the most extensively validated and accepted RA composite measure. However, patient-reported outcomes and tender/swollen joint counts exhibit significant intra- and interassessor variability and can, for example, be influenced by cumulative damage and comorbidities. In principle, biomarkers have the potential to provide objective measurements of the disease processes underlying RA, although common laboratory tests such as erythrocyte sedimentation rate (ESR) and C-reactive protein (CRP) are nonspecific markers of inflammation and can be unexpectedly low or even normal in RA patients with active disease. ESR and CRP measurement may therefore not be useful in all RA patients, and a comprehensive panel of relevant serum protein biomarkers reflecting and covering as much as possible the different interactions in RA (including cytokines, adhesion molecules, growth factors, matrix metalloproteinases, and hormones) has more potential to assess DA in a specific and objective manner encompassing all different features of the disease beyond systemic inflammation. A novel Vectra DA test (Crescendo Bioscience) has been developed and recently commercialized to determine an objective multibiomarker disease activity (MBDA) score, employing an algorithm to combine the levels of 12 serum biomarkers into a single score from 1 to 100 [68] (Table 3). A score below 25 indicates molecular remission indicative of true disease quiescence, while the score further categorizes RA into low (>25 to ≤29), moderate (>29 to ≤44), or high (>44) DA [68]. During development of the test, biomarker assays from several different platforms were used in feasibility studies to identify biomarkers of potential significance. These assays were adapted to a multiplex platform for training and validation of the algorithm against DAS28 [68]. The MBDA algorithm has subsequently been evaluated and validated in independent cohorts and tight controlled studies such as CAMERA [69], BeSt [70], and Nested-1 [71]. These studies demonstrated that the MBDA score is significantly associated with the DAS28-CRP score and other clinical measures [69–71] could track changes in DA over time [70,71], could indicate treatment responses [69], and could discriminate clinical responders from nonresponders [70,71]. Ongoing and future research will further evaluate the ability of the score to indicate risk of critical patient outcomes such as RA flares, joint damage progression, and disability. In conclusion, the MBDA test is by design an exclusively biomarker-based assessment. While this gives certain advantages, including efficiency and objectivity, it does not include physical examination and should not be seen as a replacement for the existing symptom-based DA measures, but rather as a complementary tool to provide objective, quantitative data to help patient management and clinical decision making in RA. The MBDA score could, for example, be used to enable monthly monitoring of DA while allowing clinical assessment to take place less frequently. Moreover, it could be used to support management of difficult cases such as patients with comorbidities or conflicting physician versus patient assessment [68].

From Describing the Status Quo to Predicting/Modeling the (yet) Unknown

Modeling and simulation is a powerful in silico tool to make predictions and to formulate hypotheses based on available data. In the past, this technique has mainly been used to predict the pharmacokinetic (PK) behavior of a drug. More recently PD and even efficacy and safety end points were also used in such models and many of those parameters are biomarkers. In the following, a couple of examples are being presented illustrating the power of modeling using biomarkers and the potential applications. The examples range from classical PK/PD modeling, over the modeling of cardiac safety biomarkers or different in silico models of disease to even more complex models involving PK, PD, and disease biomarkers.

PK modeling can provide an estimate of the PK behavior of a drug. When this is combined with PD biomarkers this becomes much more powerful. PK/PD modeling is regularly used with biological drugs for which the risk of immunogenicity exists. An immunogenic reaction can give rise to neutralizing antibodies against the drug, and this may be readily recognized by a mismatch between PK and PD behavior of a drug. Also, for small molecules this can yield important information. For instance, Landersdorfer and colleagues [73] modeled the PK of a dipeptidyl peptidase 4 inhibitor vildagliptin with its PD effect, the inhibition of the breakdown of GLP-1. The modeling correctly predicted the drug target to be also a drug-metabolizing enzyme, i.e., a target-mediated drug disposition could be deduced for this compound.

Many drug candidates fail during development based on safety issues. Often, these safety findings are only identified in large-scale clinical trials or even after marketing approval. Many attempts have therefore been made to front load drug development efforts with dedicated safety assessments. One prominent example is cardiovascular safety assessment and more precisely the prediction of cardiac arrhythmias like torsade de pointes by in vitro testing of hERG channel binding or in vivo measurement of the QT interval. However, there is no clear correlation between interaction of a compound with the hERG channel and the induction of cardiac arrhythmias: while many drugs that induce torsade de pointes interact with the hERG channel, there are also many compounds that bind to hERG but do not cause any arrhythmias. Thus while early testing for hERG channel binding might have prevented a few drugs that cause arrhythmias to enter clinical development, it probably might have caused more potential valuable therapies to be abandoned for no reason. In an attempt to enhance the predictive value of in vitro testing and early clinical ECG assessments, computer models have been proposed to integrate the different cardiac biomarkers in order to more precisely predict the cardiac risk of a drug candidate [74]. The available models are able to describe, for instance, the cardiac action potential for different animal species. Rabbits and guinea pigs are the most sensitive and predictive species for the drug-induced QT prolongation and these models correctly calculated the prolongation by a certain degree of hERG channel inhibition. Moreover, for each of the commonly assessed cardiac biomarkers (e.g., action potential duration), these models calculate which ion permeabilities contribute most prominently to each biomarker, thereby helping the interpretation of drug effects and their significance. Even more important is the translational value of such models. The cardiac action potential exhibits large species differences and it is important to understand whether a specific effect seen in one nonclinical species translates into a clinically meaningful effect. These in silico models can make predictions as to the potential effects in men and can thus help to design specific trials to assess the hypothesis.

The pathophysiology of many diseases is not well understood, thereby limiting the rational design of new therapies. Many especially chronic diseases are caused by a disturbance of a number of complex physiological systems. Biomarker research has often taken the reductionistic approach, e.g., by genome-wide association studies to identify single nucleotide polymorphism that correlates with a certain condition. This research has identified, for instance, hundreds of genes, mRNAs, proteins, and metabolites that may be correlated with renal dysfunction [75]. Only recently, efforts have been initiated to evaluate those single biomarkers in a systems biology approach such as the Syskid project [76]. This project uses a systems biology approach to integrate information on renal biomarkers to enable the early diagnosis of chronic kidney diseases, the interaction with other chronic diseases such as diabetes, and to enable more effective treatment.

Drug testing in animal models and clinical trials is an expensive and cumbersome endeavor. In silico models have been developed for a couple of diseases in order to help select the most promising drug targets, drug combinations, and biomarkers to advance only the most promising drugs to (pre)clinical testing. Jackson and Radivoyevitch have created an in silico model of neutrophil overactivation to model the production of reactive oxygen species in chronic inflammatory lung diseases [77]. Key biomarkers were selected that represent the main signaling pathways in neutrophils leading to the release of reactive oxygen species. By applying this model to a number of candidate drugs (or mechanisms) they predicted a combination of two drugs to provide synergistic effects on reactive oxygen production with a fast onset and long duration of action. Those combinations remain, however, to be clinically evaluated.

Even further advanced is the approach whereby disease modeling is combined with PK/PD modeling. Dwivedi et al. have combined a model of interleukin-6 (IL-6) and IL-6 receptor (IL-6R) signaling with a PK/PD model for the anti-IL-6 receptor antibody tocilizumab [78]. The model was successfully validated in a clinical trial for Crohn disease to predict plasma levels of the inflammatory biomarker CRP. The model was then used to identify the most efficient way to interfere with IL-6 signaling as a potential treatment of Crohn disease, e.g., an antibody directed against IL-6 (the ligand) more or less active than the one directed against specific forms of the IL-6 receptor. The model predicted that the inhibition of IL-6 interaction with both of its receptors, the soluble and the membrane bound, would be most effective. This hypothesis needs, however, further clinical validation.

Modeling of biomarkers can also be used to make treatment decisions, e.g., in antiviral therapy. Vergu et al. (2004) have created a mathematical model to predict treatment response to combinations of antiretroviral drugs based on baseline levels of T-cell subsets (CD4 and CD8) as well as genotypes of the HIV strains present in the respective patient [79]. While the analysis of the single biomarkers (i.e., the T-cell subsets and the HIV strain genotypes) give some indication as to which drugs may provide benefit, it is the modeling of the interaction between the single markers that is much more powerful especially in cases of initial treatment resistance.

These examples provide an excellent illustration for the need of forward and backward translation between in vitro or preclinical models, in silico models, and clinical trials and how biomarkers can be instrumental in these translational processes.

BIOMARKER DOCUMENTATION

From Isolated Reporting to Meaningful Representation of Results that Allows Replication and Confirmation

Although many scientific publications report biomarker results, for only a few of those biomarkers further validation was demonstrated which is needed to ultimately allow their clinical use. One of the reasons that so many biomarkers fail down the road is that the description of the methodology used and of the results generated does not allow replication of the results nor proper evaluation of the significance of the findings. More transparent and complete reporting of biomarker results would therefore facilitate sound interpretation, application, and replication of study results and would enable others to assess the utility of these biomarkers.

One initiative to improve the reporting of biomarkers is the "reporting recommendations for tumor marker prognostic studies" (REMARK), which was designed to help authors ensure that reports of their tumor marker studies contain the information that readers need. The REMARK checklist consists of 20 items that should be taken into account when publishing results of tumor marker prognostic studies [80]. These recommendations are focusing on tumor materials, such as tumor sections or DNA- and RNA-based assays, but a lot of these items are also relevant for biomarker reporting in general and REMARK can easily apply to other diseases in addition to cancer. The processes of measuring and analyzing the biomarker results may differ, but some of the same study reporting principles apply as will be explained in the next paragraphs and are summarized in Table 4.

Foremost, it is for every reported biomarker important to clearly provide early in the study report the marker examined, the study objectives, and the prespecified hypotheses [80]. A thorough description of the marker is necessary for an understanding of the biological rationale and potential clinical application. Prespecified hypotheses are based on prior research or systematic review of the literature, and they are stated before the study is initiated. The objectives are mostly linked with these hypotheses and should ideally be formulated in terms of measures that can be statistically evaluated.

TABLE 4   Selected Recommendations from the REMARK Checklist Which Were Discussed Here and Which Are Deemed Suitable for All Biomarkers

---

INTRODUCTION AND DISCUSSION

- State the marker examined, the study objectives, and any prespecified hypotheses (*REMARK recommendation 1*)
- Interpret the results in the context of the prespecified hypotheses and other relevant studies; include a discussion of limitations of the study (*REMARK recommendation 19*)

SPECIMEN CHARACTERISTICS

- Describe type of biological material used (including control samples) and methods of preservation and storage (*REMARK recommendation 4*)

ASSAY METHODS

- Specify the assay method used and provide (or reference) a detailed protocol, including specific reagents or kits used, quality control procedures, reproducibility assessments, quantitation methods, and scoring and reporting protocols. Specify whether and how assays were performed blinded to the study end point. (*REMARK recommendation 5*)

STATISTICAL ANALYSIS METHODS

- Specify all statistical methods, including details of any variable selection procedures and other model-building issues, how model assumptions were verified, and how missing data were handled. (*REMARK recommendation 10*)
- Clarify how marker values were handled in the analyses. (*REMARK recommendation 11*)

Adapted from Ref. [80].

Second, it is for every biomarker also important to describe in detail the type and source of the biological biospecimen or samples used [80]. As much information as possible about the timing of sample collection and the way the samples were collected, processed, and stored should be included in the report. The storage conditions not only entail how but also how long samples were stored prior to performing the biomarker assay. Much has been written about the potential confounding effects of pre-analytical handling of samples and biological materials (see also Sections From Blood- to Tissue-Specific Biomarkers and From Static to Functional Biomarker Assays), and several organizations have recently published articles addressing best practices for sample handling, including the Biospecimen Reporting for Improved Study Quality guidelines [81]. These guidelines provide comprehensive recommendations for what information should be reported regarding specimen characteristics and methods of specimen processing and handling.

Assay methods should be reported in a complete and transparent way with a level of detail that would enable another laboratory to reproduce the protocol [80]. It has been demonstrated for many markers that a slightly different methodology can produce systematically different results. If another widely accessible document (such as a published paper) details the exact assay method used, it is acceptable to reference that other document without repeating all the technical details. If a commercially available kit is used for the assay, it is important to state whether the kit instructions were followed exactly; any deviations from the kit's recommended procedures must be fully explained in the report. The results of the assay characterization and qualification experiments should be reported to provide a sense of the dynamic range and performance (precision and accuracy) of the assay, even if the assay itself has previously been characterized by the manufacturer or in another laboratory. Stability of the biomarker, spike recovery in the biological matrix of interest, and assay parallelism should also be assessed [82]. It is important to report any procedures, such as use of spiked quality control samples or sample controls with endogenous biomarker levels, that are employed for run acceptance or to assess consistency of assay results over time [82]. For assays in more advanced state of development, additional examples could include the bridging strategy for new lots of reagents or instrument calibration procedures.

Regarding data evaluation, a substantial improvement of reporting of the used statistical methods is needed and specification of all statistical methods should be clearly performed in the report in such a way that the knowledgeable reader with access to the original data would be able to verify the reported results [80]. Some assessment of the data quality usually takes place prior to the main statistical analyses of the data, and data values may be changed or removed if they are deemed unreliable. These manipulations and preanalysis decisions could have a substantial impact on the results and should also be reported. Examples are omitting extreme outliers, use of transformations or normalizations, recategorization for categorical variables, etc. The analysis may further include an examination of the relationship of the biomarker to other variables from the study such as clinical outcome. Methods for summarizing associations with other variables should therefore also be described and it is of course important to report whether biomarker assessments were made blinded to clinical outcome.

Last but not least, selective reporting makes the evaluation of the usefulness of a biomarker also more complex [80]. Publication of selective findings from a particular study will lead to bias when publication choices are made with the knowledge of study findings. Selection is mostly in relation to whether or not results were statistically significant ($p < 0.05$) or show a trend in the favored direction. Evidence of biased nonpublication of whole studies has been accumulating over many years, but recently research has demonstrated evidence of additional within-study selective reporting [83,84]. Obviously, not only selective reporting by authors but also the preference of some scientific journals for publishing only positive results is an important hurdle to the reliable assessment of the clinical utility of a biomarker.

PROMISING APPLICATIONS

From the Discovery of Pathomechanisms to the Definition of Patient (Sub)populations

Stratification of patients has been one of the great promises of the twenty-first century and entails matching a patient with a particular therapy most likely to be effective and safe based on biomarkers specific for that type of responders ("personalized medicine or stratified medicine" [85]). Stratification biomarkers are objectively measured characteristics that preferentially capture a causal relationship between the presence of the candidate biomarker in the patient and the clinical outcome to the particular therapy. While biomarker discovery is very actively pursued, the use of biomarkers in clinical trials and clinical practice is still in its infancy: only a small number of cancer biomarkers are yearly being approved for use by the FDA and the European Medicines Agency [86]. While the usefulness of stratification biomarkers has long been evident, the hurdles and challenges are becoming more tangible as well.

Oncology is without any doubt one of the prime areas of targeted therapies, which has evolved from the knowledge of the mutations underlying the disease. Whereas in 2006, the number of prominent examples of personalized medicine treatments and diagnostics has increased to 13 products (69% of which were for cancer), by 2014, the number of approved combinations had already increased to 113 [87,88]. Overall, extension of this approach to polygenic disorders, such as RA or psoriatic arthritis remains challenging as biomarkers need to be identified that are causal instead of bystanders and as in these disorders the underlying pathways are diverse and complex.

In RA, therapies are currently still being selected on a trial-and-error basis and overall in less than 50% of patients, a more than 50% improvement in the overall disease status in response to biological disease-modifying antirheumatic drugs (bDMARDS) is reported [85]. Initiatives like MATURA are being undertaken to elucidate how patients will respond to bDMARDS, exploring multiple platforms like ultrasound imaging, and psychosocial and genetic factors. Some progress has been made in identifying subpopulations in RA based on gene expression of synovial fibroblasts; patients could be segregated in more myeloid or more lymphocytic subpopulations. Soluble serum biomarkers were identified that could differentiate between these phenotypes: a high baseline of serum soluble intercellular adhesion molecule 1 was associated with the myeloid phenotype, whereas a high baseline of serum C-X-C motif chemokine 13 (CXCL13) was associated with the lymphoid phenotype. A differential relationship with the clinical response to anti-TNFα compared with anti-IL6R treatment was described for these phenotypes: patients with the myeloid phenotype exhibited the strongest response to anti-TNFα therapy, whereas those with lymphocytic phenotype demonstrated a more robust response to treatment with an anti-IL6R antibody [89,90]. Another example of progress in the field of stratification in RA is the description of a 24-biomarker signature which might be informative of responsiveness to anti-TNFα therapy [91], notwithstanding that presently no predictive biomarkers exist on the individual level which can be used to support decision making for individual patients in RA.

The advantages for patient stratification are numerous and compelling:

1. Identification of patient subpopulations, as most diseases are heterogeneous in symptom range and intensity, in underlying causal pathways, and responsiveness to different drugs
2. Creating a match between the patient and the most effective treatment
3. Providing effective treatment that might prevent irreversible damage (e.g., in case of an arthritic disease) as less time is lost looking for this match
4. The increase of safety by avoiding side effects (e.g., avoidance of the antifolate drug MTX in certain folate-dependent enzyme genotypes, at risk of experiencing MTX toxicity [92]) in patients that are possibly even without a therapeutic benefit (e.g., bevacizumab [93])
5. The improvement of drug development by enriching trials for patients more likely to respond, which in turn will reduce trial size (prospective stratification needed) and which will increase the chance of establishing therapeutic efficacy
6. Stratification may also rescue drugs that fail in the total population or experience safety issues in subpopulations
7. Stratification should reduce total health costs by avoiding unnecessary treatment

Regardless of the great potential of stratification, it comes with challenges and requires effort, time, and cost and should be well planned within drug development and will have great impact on clinical trial design:

1. Finding stratification biomarkers requires thorough understanding of the disease and the pathways involved: both the target and alternative pathways that influence the effect of the drug. In addition, many diseases are multifactorial syndromes, made up of different disease subtypes with different phenotypes and genotypes, rendering stratification biomarker identification extremely difficult. Also in oncology, the guidance for treatment decisions is shifting by the acknowledgment of the impact of both tumor and nontumor cells of different types, the extracellular matrix, and responsiveness/ resistance to other therapies. Hence, it is becoming more complex than the single gene or mutation-driven prediction for the early

stratification examples, trastuzumab or Gleevec (see Section From Peptide Molecules to Nonbiochemical Biomarkers for the use of biomarker panels).

2. In heterogeneous diseases like RA and SLE, the outcome measures to demonstrate the effect in a certain subpopulation are more comprehensive than in oncology, where mostly survival and remission are evaluated. It is thus more difficult to demonstrate significant outcomes.

3. Trials need to be designed to yield convincing evidence, proving effect specifically in a certain patient population. In this context, retrospective data are of value, although further confirmation is needed in prospective trials.

4. Population stratification narrows the market share for the drug by eliminating otherwise prospective patients. On the other hand, one potential is that populations become larger since the same mechanisms may be relevant in other diseases. For instance, a certain oncogenic pathway that could be targeted by a particular drug may be the cause of tumors in different organs.

5. For many stratification biomarkers, a threshold will be required as a yes/no answer is not applicable. Setting this threshold is difficult and will have implications ethically (excluding patients from treatment), economically (downsizing patient population), and on efficacy (the larger the treatable range, the more heterogeneous the patient population might be).

6. Not only the biomarker concept but also the assay needs validation. For validation of an analytical method, significant guidance and standardization is in place. However, also standardized specimen acquisition, processing and storage, and appropriate postanalytical methods need further optimization and standardization (see Section From Isolated Reporting to Meaningful Representation of Results that Allows Replication and Confirmation'). If no validated assay is clinically available, a companion diagnostic will have to be developed. Development of the drug and the diagnostic is preferably simultaneous, although both develop under guidance of different regulatory bodies and with different time lines.

7. Even if a test is available, the market may be reluctant to immediately accept it and this willingness will differ greatly between countries.

How to go about patient stratification? Evidently the best candidates to define disease subtypes are mechanistic biomarkers directly involved in the pathogenesis of a disease. Consequently, if the patient responds to treatment the disease was (partly) caused by the target (e.g., global gene expression analysis of synovial tissue in RA: highlighting the heterogeneity in RA synovium and defining a more lymphocytic versus myeloid-oriented pathology). High-throughput omics-based technologies have been developed (genomics, proteomics, metabolomics, and microRNA) that generate a wealth of information but the difficulty remains how to deduce relevant information [94]. Sensitive, specific, and predictive biomarkers are required for clinical decision making. Proteomics takes into account both expression and posttranslational modifications but is limited due to biased screening and technical challenges. To interpret the information generated with these platforms, information extraction is important. This is still mainly a manual and thus slow process, although improvements are being made by intensified data sharing and uniformity (of, e.g., nomenclature [95]) or modeling approaches (see Section From Single Biomarkers to Biomarker Patterns). Overall it has become clear that single stratification biomarkers (e.g., overexpression of the *HER2* gene indicates the high likelihood that a patient with breast cancer will respond to trastuzumab) will be more the exception and that the limited predictive power of single biomarkers can be improved by extension to a panel (see Section From Single Biomarkers to Biomarker Patterns), better describing the responses and outcomes and increasing sensitivity. Validation of the concept of a stratification biomarker is exerted by confirmation in prospective clinical trials and by demonstration of the impact on disease management and health care costs.

In summary, the theoretical advantages of stratification are most compelling and many clinical examples have confirmed these theoretical benefits. Progress has been made in facilitating the knowledge sharing, biomarker validation, diverse biomarker platforms, regulatory guidance, and clinical trial design and this will further pave the way for future development. Stratification, however, also imposes extra hurdles or complexity for development, which can at least partially be circumvented by considering the need and the strategy early on during drug development.

CONCLUSIONS

To conclude we advocate a number of recommendations for the use of biomarkers in a translational strategy:

First, one should think backward with the target in mind (e.g., a drug target product profile) and translate the wish list into specific biomarkers and use those early on.

Second, one should think outside the classical lines when considering biomarkers:

- Images tell more than 1000 words: Biomarker imaging is an elegant and powerful way to provide in vivo real-time information and limited invasiveness.

- Go local: while the blood stream is an easily accessible compartment to sample for biomarkers, the target tissue of interest may provide more undiluted information on the biomarkers of interest.
- Function matters: Functional biomarkers may be more sensitive markers of a certain condition than static markers.
- Think broad: Biomarkers can include not only proteins but also more complex constructs such as a quality-of-life measurement.

Third, the interpretation of biomarker results can be enhanced by considering biomarker patterns and by putting the data into algorithms or a computational model. Model results may help to explain the unexpected and refine the initial hypothesis.

Next, reporting of the results (even though the original hypothesis may have been disproven) should encourage others to reproduce and further build on the progress.

Finally, one should realize that "less can be more," i.e., refining your target patient population may yield better effects. Think about stratification early on in the development, while thoroughly investigating the role of the target in a certain disease and pathways affecting it, as well as the stratification biomarker that gives an indication which patients may benefit most.

References

[1] Biomarkers and surrogate endpoints: preferred definitions and conceptual framework. Clin Pharmacol Ther 2001;69(3):89–95.
[2] Pedersen KO. Ultracentrifugal and electrophoretic studies on fetuin. J Phys Colloid Chem 1947;51(1):164–71.
[3] van Kooten C, Banchereau J. CD40-CD40 ligand. J Leukoc Biol 2000;67(1):2–17.
[4] Boumpas DT, et al. A short course of BG9588 (anti-CD40 ligand antibody) improves serologic activity and decreases hematuria in patients with proliferative lupus glomerulonephritis. Arthritis Rheum 2003;48(3):719–27.
[5] Couzin J. Drug discovery. Magnificent obsession. Science 2005;307(5716):1712–5.
[6] Langer F, et al. The role of CD40 in CD40L- and antibody-mediated platelet activation. Throm Haemostasis 2005;93(6):1137–46.
[7] Robles-Carrillo L, et al. Anti-CD40L immune complexes potently activate platelets in vitro and cause thrombosis in FCGR2A transgenic mice. J Immunol 2010;185(3):1577–83.
[8] Mirabet M, et al. Platelet pro-aggregatory effects of CD40L monoclonal antibody. Mol Immunol 2008;45(4):937–44.
[9] Deambrosis I, et al. Inhibition of CD40-CD154 costimulatory pathway by a cyclic peptide targeting CD154. J Mol Med 2009;87(2):181–97.
[10] Wakefield I, et al. An assessment of the thromboembolic potential of CDP7657, a monovalent Fab' PEG anti-CD40L antibody, in Rhesus macaques. Arthritis Rheum 2010;62.
[11] Tocoian A, et al. First-in-human trial of the safety, pharmacokinetics and immunogenicity of a PEGylated anti-CD40L antibody fragment (CDP7657) in healthy individuals and patients with systemic lupus erythematosus. Lupus 2015.
[12] Cary KC, Cooperberg MR. Biomarkers in prostate cancer surveillance and screening: past, present, and future. Ther Adv Urol 2013;5(6):318–29.
[13] Martin SK, et al. Emerging biomarkers of prostate cancer (Review). Oncol Rep 2012; 28(2):409–17.

[14] McGregor M, et al. Screening for prostate cancer: estimating the magnitude of overdetection. CMAJ 1998;159(11):1368–72.

[15] Andriole GL, et al. Mortality results from a randomized prostate-cancer screening trial. N Engl J Med 2009;360(13):1310–9.

[16] Schroder FH, et al. Screening and prostate-cancer mortality in a randomized European study. N Engl J Med 2009;360(13):1320–8.

[17] Schroder FH, et al. Screening and prostate cancer mortality: results of the European Randomised Study of Screening for Prostate Cancer (ERSPC) at 13 years of follow-up. Lancet 2014;384(9959):2027–35.

[18] Moyer VA. Screening for prostate cancer: U.S. Preventive Services Task Force recommendation statement. Ann Intern Med 2012;157(2):120–34.

[19] Gu Z, et al. Prostate stem cell antigen (PSCA) expression increases with high Gleason score, advanced stage and bone metastasis in prostate cancer. Oncogene 2000; 19(10):1288–96.

[20] Saeki N, et al. Prostate stem cell antigen: a Jekyll and Hyde molecule? Clin Cancer Res 2010;16(14):3533–8.

[21] Zhao Z, et al. Prostate stem cell antigen mRNA expression in preoperatively negative biopsy specimens predicts subsequent cancer after transurethral resection of the prostate for benign prostatic hyperplasia. Prostate 2009;69(12):1292–302.

[22] Thomas-Kaskel AK, et al. Vaccination of advanced prostate cancer patients with PSCA and PSA peptide-loaded dendritic cells induces DTH responses that correlate with superior overall survival. Int J Cancer 2006;119(10):2428–34.

[23] Galon J, et al. Cancer classification using the Immunoscore: a worldwide task force. J Transl Med 2012;10:205.

[24] Pages F, et al. Immune infiltration in human tumors: a prognostic factor that should not be ignored. Oncogene 2010;29(8):1093–102.

[25] Paniccia R, et al. Platelet function tests: a comparative review. Vasc Health Risk Manag 2015;11:133–48.

[26] Born GV. Aggregation of blood platelets by adenosine diphosphate and its reversal. Nature 1962;194:927–9.

[27] Femia EA, et al. Comparison of different procedures to prepare platelet-rich plasma for studies of platelet aggregation by light transmission aggregometry. Platelets 2012;23(1): 7–10.

[28] Toth O, et al. Multiple electrode aggregometry: a new device to measure platelet aggregation in whole blood. Thromb Haemostasis 2006;96(6):781–8.

[29] Sibbing D, et al. Clopidogrel response status assessed with multiplate point-of-care analysis and the incidence and timing of stent thrombosis over six months following coronary stenting. Thromb Haemostasis 2010;103(1):151–9.

[30] Ranucci M, et al. Multiple electrode whole-blood aggregometry and bleeding in cardiac surgery patients receiving thienopyridines. Ann Thorac Surg 2011;91(1):123–9.

[31] Freer G, Rindi L. Intracellular cytokine detection by fluorescence-activated flow cytometry: basic principles and recent advances. Methods 2013;61(1):30–8.

[32] De Rosa SC. Vaccine applications of flow cytometry. Methods 2012;57(3):383–91.

[33] Bull M, et al. Defining blood processing parameters for optimal detection of cryopreserved antigen-specific responses for HIV vaccine trials. J Immunol Methods 2007; 322(1–2):57–69.

[34] Goepfert PA, et al. Phase 1 safety and immunogenicity testing of DNA and recombinant modified vaccinia Ankara vaccines expressing HIV-1 virus-like particles. J Infect Dis 2011;203(5):610–9.

[35] Harrer T, et al. Safety and immunogenicity of an adjuvanted protein therapeutic HIV-1 vaccine in subjects with HIV-1 infection: a randomised placebo-controlled study. Vaccine 2014;32(22):2657–65.

[36] Keefer MC, et al. A phase I trial of preventive HIV vaccination with heterologous poxviral-vectors containing matching HIV-1 inserts in healthy HIV-uninfected subjects. Vaccine 2011;29(10):1948–58.

[37] Buchbinder SP, et al. Efficacy assessment of a cell-mediated immunity HIV-1 vaccine (the Step Study): a double-blind, randomised, placebo-controlled, test-of-concept trial. Lancet 2008;372(9653):1881–93.

[38] McElrath MJ, et al. HIV-1 vaccine-induced immunity in the test-of-concept Step Study: a case-cohort analysis. Lancet 2008;372(9653):1894–905.

[39] Petrella JR. Neuroimaging and the search for a cure for Alzheimer disease. Radiology 2013;269(3):671–91.

[40] Pimplikar SW, et al. Amyloid-independent mechanisms in Alzheimer's disease pathogenesis. J Neurosci 2010;30(45):14946–54.

[41] Chetelat G. Alzheimer disease: Abeta-independent processes-rethinking preclinical AD. Nat Rev Neurol 2013;9(3):123–4.

[42] Perani D. Functional neuroimaging of cognition. Handb Clin Neurol 2008;88:61–111.

[43] Kung HF, et al. 18F stilbenes and styrylpyridines for PET imaging of A beta plaques in Alzheimer's disease: a miniperspective. J Med Chem 2010;53(3):933–41.

[44] Yeo JM, Waddell B, Khan Z, Pal S. A systematic review and meta-analysis of 18F-labeled amyloid imaging in Alzheimer's disease. Alz Dementia Diagn Assess Dis Monit 2015; 1(1):5–13.

[45] Laforce Jr R, Rabinovici GD. Amyloid imaging in the differential diagnosis of dementia: review and potential clinical applications. Alz Res Ther 2011;3(6):31.

[46] Jack Jr CR, Holtzman DM. Biomarker modeling of Alzheimer's disease. Neuron 2013;80(6):1347–58.

[47] Cohen AD, Klunk WE. Early detection of Alzheimer's disease using PiB and FDG PET. Neurobiol Dis 2014;72(Pt A):117–22.

[48] Braun HJ, Gold GE. Diagnosis of osteoarthritis: imaging. Bone 2012;51(2):278–88.

[49] Emrani PS, et al. Joint space narrowing and Kellgren-Lawrence progression in knee osteoarthritis: an analytic literature synthesis. Osteoarthr Cartilage 2008;16(8):873–82.

[50] Conaghan PG, et al. Summary and recommendations of the OARSI FDA osteoarthritis Assessment of Structural Change Working Group. Osteoarthr Cartilage 2011;19(5):606–10.

[51] Wang Y, et al. Use magnetic resonance imaging to assess articular cartilage. Ther Adv Musculoskelet Dis 2012;4(2):77–97.

[52] Hunter DJ, et al. Systematic review of the concurrent and predictive validity of MRI biomarkers in OA. Osteoarthr Cartilage 2011;19(5):557–88.

[53] Kellgren JH, Lawrence JS. Radiological assessment of osteo-arthrosis. Ann Rheum Dis 1957;16(4):494–502.

[54] Altman RD, Gold GE. Atlas of individual radiographic features in osteoarthritis, revised. Osteoarthr Cartilage 2007;15(Suppl. A):A1–56.

[55] Peterfy CG, et al. Whole-organ magnetic resonance imaging score (WORMS) of the knee in osteoarthritis. Osteoarthr Cartilage 2004;12(3):177–90.

[56] Hunter DJ, et al. The reliability of a new scoring system for knee osteoarthritis MRI and the validity of bone marrow lesion assessment: BLOKS (Boston Leeds Osteoarthritis Knee Score). Ann Rheum Dis 2008;67(2):206–11.

[57] Woolf AD, Pfleger B. Burden of major musculoskeletal conditions. B World Health Organ 2003;81(9):646–56.

[58] Sofat N, Kuttapitiya A. Future directions for the management of pain in osteoarthritis. Int J Clin Rheumatol 2014;9(2):197–276.

[59] Sanders D, et al. Pharmacologic modulation of hand pain in osteoarthritis: a double-blind placebo-controlled functional magnetic resonance imaging study using naproxen. Arthritis Rheumatol 2015;67(3):741–51.

[60] Dobson F, et al. OARSI recommended performance-based tests to assess physical function in people diagnosed with hip or knee osteoarthritis. Osteoarthr Cartilage 2013;21(8):1042–52.

[61] Dobson F, et al. OARSI paper: recommended performance-based tests to assess physical function in people diagnosed with hip or knee osteoarthritis. OARSI Website; 2013.

[62] Gadducci A, Sergiampietri C, Tana R. Alternatives to risk-reducing surgery for ovarian cancer. Ann Oncol 2013;24(Suppl. 8):viii47–53.

[63] Nolen BM, Lokshin AE. Biomarker testing for ovarian cancer: clinical utility of multiplex assays. Mol Diagn Ther 2013;17(3):139–46.

[64] Visintin I, et al. Diagnostic markers for early detection of ovarian cancer. Clin Cancer Res 2008;14(4):1065–72.

[65] Buchen L. Cancer: missing the mark. Nature 2011;471(7339):428–32.

[66] Moore RG, et al. The use of multiple novel tumor biomarkers for the detection of ovarian carcinoma in patients with a pelvic mass. Gynecol Oncol 2008;108(2):402–8.

[67] Zhang Z, Chan DW. The road from discovery to clinical diagnostics: lessons learned from the first FDA-cleared in vitro diagnostic multivariate index assay of proteomic biomarkers. Cancer Epidemiol Biomarkers Prev 2010;19(12):2995–9.

[68] Centola M, et al. Development of a multi-biomarker disease activity test for rheumatoid arthritis. PLoS One 2013;8(4):e60635.

[69] Bakker MF, et al. Performance of a multi-biomarker score measuring rheumatoid arthritis disease activity in the CAMERA tight control study. Ann Rheum Dis 2012;71(10):1692–7.

[70] Hirata S, et al. A multi-biomarker score measures rheumatoid arthritis disease activity in the BeSt study. Rheumatology 2013;52(7):1202–7.

[71] Curtis JR, et al. Validation of a novel multibiomarker test to assess rheumatoid arthritis disease activity. Arthritis Care Res 2012;64(12):1794–803.

[72] Moore RG, et al. A novel multiple marker bioassay utilizing HE4 and CA125 for the prediction of ovarian cancer in patients with a pelvic mass. Gynecol Oncol 2009;112(1):40–6.

[73] Landersdorfer CB, He YL, Jusko WJ. Mechanism-based population pharmacokinetic modelling in diabetes: vildagliptin as a tight binding inhibitor and substrate of dipeptidyl peptidase IV. Br J Clin Pharmacol 2012;73(3):391–401.

[74] Corrias A, et al. Arrhythmic risk biomarkers for the assessment of drug cardiotoxicity: from experiments to computer simulations. Philos Trans Ser A 2010;368(1921):3001–25.

[75] Pesce F, Pathan S, Schena FP. From -omics to personalized medicine in nephrology: integration is the key. Nephrol Dialysis Transplant 2013;28(1):24–8.

[76] Heinzel A, et al. From molecular signatures to predictive biomarkers: modeling disease pathophysiology and drug mechanism of action. Front Cell Dev Biol 2014;2:37.

[77] Jackson RC, Radivoyevitch T. Modelling c-Abl signalling in activated neutrophils: the anti-inflammatory effect of seliciclib. BioDiscovery 2013;7(4):4.

[78] Dwivedi G, et al. A multiscale model of interleukin-6-mediated immune regulation in Crohn's disease and its application in drug discovery and development. CPT Pharmacometrics Syst Pharmacol 2014;3:e89.

[79] Vergu E, Mallet A, Golmard JL. Available clinical markers of treatment outcome integrated in mathematical models to guide therapy in HIV infection. J Antimicrob Chemother 2004;53(2):140–3.

[80] Altman DG, et al. Reporting recommendations for tumor marker prognostic studies (REMARK): explanation and elaboration. BMC Med 2012;10:51.

[81] Moore HM, et al. Biospecimen reporting for improved study quality. Biopreservation Biobanking 2011;9(1):57–70.

[82] Lee JW, et al. Fit-for-purpose method development and validation for successful biomarker measurement. Pharm Res 2006;23(2):312–28.

[83] Dwan K, et al. Systematic review of the empirical evidence of study publication bias and outcome reporting bias. PLoS One 2008;3(8):e3081.

[84] Williamson PR, et al. Outcome selection bias in meta-analysis. Stat Methods Med Res 2005;14(5):515–24.

[85] Lindstrom TM, Robinson WH. Biomarkers for rheumatoid arthritis: making it personal. Scand J Clin Lab Inv Suppl 2010;242:79–84.

[86] Taube SE, et al. A perspective on challenges and issues in biomarker development and drug and biomarker codevelopment. J Natl Cancer Inst 2009;101(21):1453–63.

[87] Singh V. Companion diagnostics poised for a breakout. The Burrill Report [webpage on the Internet] 2012;2(9).

[88] The case for personalized medicine. 1st ed. (2006) and 4th ed. (2014). Available from: www.personalizedmedicinecoalition.org.

[89] Galligan CL, et al. Distinctive gene expression signatures in rheumatoid arthritis synovial tissue fibroblast cells: correlates with disease activity. Genes Immun 2007;8(6):480–91.

[90] Dennis Jr G, et al. Synovial phenotypes in rheumatoid arthritis correlate with response to biologic therapeutics. Arthritis Res Ther 2014;16(2):R90.

[91] Hueber W, et al. Blood autoantibody and cytokine profiles predict response to anti-tumor necrosis factor therapy in rheumatoid arthritis. Arthritis Res Ther 2009;11(3):R76.

[92] Weisman MH, et al. Risk genotypes in folate-dependent enzymes and their association with methotrexate-related side effects in rheumatoid arthritis. Arthritis Rheum 2006;54(2):607–12.

[93] Afranie-Sakyi JA, Klement GL. The toxicity of anti-VEGF agents when coupled with standard chemotherapeutics. Cancer Lett 2015;357(1):1–7.

[94] Castro-Santos P, Laborde CM, Diaz-Pena R. Genomics, proteomics and metabolomics: their emerging roles in the discovery and validation of rheumatoid arthritis biomarkers. Clin Exp Rheumatol 2015;33(2):279–86.

[95] Deyati A, et al. Challenges and opportunities for oncology biomarker discovery. Drug Discov Today 2013;18(13–14):614–24.

[96] Jain KK. The Handbook of biomarkers. Berlin: Springer; 2010.

[97] Goodsaid F, Mattes WB. The path from biomarker discovery to regulatory qualification. Amsterdam: Elsevier; 2013.

[98] Vaidya VS, Bonventre JV. Biomarkers in medicine, drug discovery, and environmental health. Wiley; 2010.

Advancements in Data Management and Data Mining Approaches

Dimitris Kalaitzopoulos[1], Ketan Patel[1], Erfan Younesi[2]

[1]Oracle UK, Health Sciences Global Business Unit, Reading, UK;
[2]Fraunhofer Institute for Algorithms and Scientific Computing, Bioinformatics Department, Schloss Birlinghoven, Sankt Augustin, Germany

OUTLINE

Translational Medicine: Tools and Techniques
http://dx.doi.org/10.1016/B978-0-12-803460-6.00002-7

35

INTRODUCTION

The delivery of health care has reached an inflection point. There is a growing need to prevent, diagnose, and treat disease in a world where the population is aging, while keeping health expenditure at a minimum. The field of translational medicine plays a crucial role in health sciences' dramatic transformation. It allows more efficient research and development of drugs and other therapeutics, thus enabling more targeted diagnostics and therapies at the point of care.

During the translational medicine life cycle a vast amount of data are being generated. Phenotype data, such as demographics, diagnoses, lab tests, procedures, medications, as well as sample (specimen) data are collected during clinical trials and hospital encounters. The advent of molecular technologies, including next-generation sequencing (NGS), is transforming the landscape of health care and life sciences by bringing these disruptive technologies from a research setting to the clinic. Consequently, this is leading to the generation of terabytes of omics data, such as genomics, transcriptomics, proteomics, and metabolomics. Clearly there is a deluge of data that poses many challenges. Translational bioinformatics has emerged as the leading solution for management and integration of such heterogeneous, large data sets. The aim of translational bioinformatics is to develop "storage, analytic, and interpretive methods to optimize the transformation of increasingly voluminous biomedical data, and genomic data, into proactive, predictive, preventive, and participatory health" [1]. One of the missions of translational bioinformatics is to establish a way to manage the large amount of available biomedical data, with the ultimate goal of improving the health care outcome by linking molecule-, tissue-, patient-, and population-level data sets during the life cycle of translation.

In this chapter, advancements in data management and data mining approaches will be discussed, including challenges, best practices, and a sampling of tools and platforms utilized in academia and industry.

CHALLENGES OF MANAGING AND MINING BIG BIOMEDICAL DATA

Advanced laboratory techniques, on one hand, and progress in clinical diagnosis technologies, on the other hand, have led to an unprecedented increase in real-time production of data that are big in terms of volume (large amount), variety (different types/sources), velocity (massive output), variability (high inconsistencies), veracity (great range of qualities), and complexity (substantial interconnections). For example, as automated genome sequencers become cheaper, even small labs can turn into generators of big data on their own. Since these data sets are generated at high velocities, the time in which the data must be captured, analyzed, and shared become even shorter. For instance, in the United States alone, an average of 27 million medical tests are preformed per day and more than 10 million prescriptions are filled every day [2]. Here, the challenge is the storage of increasing data volumes and the processing power needed to analyze them. Cost-effective NGS technologies have led to the accelerated generation of big data, hence data management platforms need to be able to scale to support the need of the user community that want to perform analyses for interpreting data.

Unfortunately, such data sets are often heterogeneous. They are coming from a great variety of sources, both structured and unstructured. Despite structured data that are tagged and stored in data warehouses in an organized manner, unstructured data do not have a defined data model or schema and, as such, are difficult to analyze. The majority of biomedical big data is unstructured and often exists in the form of text (e.g., in biomedical literature, health records, patient blogs) or images (e.g., brain scans, microscopy images) or other types of records (e.g., audio or video streams). Moreover, such data sets suffer from a great deal of variability in terms of biases, noises, or time inconsistencies. Another major concern is the multiplicity of data formats that exist and the need to standardize data. Data standardization for both clinical and omics data is critical for facilitating integration, data analysis, and cross-study comparisons. This has created a mundane data preprocessing task for biologists and bioinformaticians to ensure that the data are clean, error-free, well-annotated, and readily usable.

Given the grand scale and rise of data, there is a need for new ways of handling big data beyond the traditional store and analysis methods. As far as the analysis itself is concerned, there is a growing field of advanced analytics and data mining techniques to address the changing needs of translational medicine. Data need to be analyzed and published, presented to conferences, and shared with collaborators and the wider community, while still preserving data security, patient privacy and ensuring regulatory compliance. Figure 1 illustrates a unified view of big data processing and challenges associated to each step.

FIGURE 1   A schematic representation of biomedical data mining and processing.

DATA MANAGEMENT FRAMEWORK

In order for the above challenges to be addressed effectively and enable researchers to undertake translational medicine approaches, a well-thought and effective data management framework should be in place. Data management is a process consisting of the planning and execution of policies, practices, and projects that acquire, collect, control, protect, share, and enhance the value of data and ensure the integrity of data assets. Data management frameworks aim at improving data management efficiency and cost-effectiveness, embedding compliance and rules into data management, achieving consistency across systems and applications, enabling growth and change easily, reducing data administration effort and costs, assisting in the selection and implementation of appropriate data management solutions, and implementing a technology-independent data architecture. When developing data and knowledge management frameworks for translational research, the following best practices should be considered:

Data Integration

Translational medicine is a multidisciplinary field, so different departments within an organization (e.g., pharmaceutical companies for drug discovery and development) need to collaborate and share data. In the new science economy model, frequently collaborations take place across organizations as is illustrated by public–private partnerships between the large academic medical centers and global pharmaceutical companies. Therefore, translational medicine data reside in disparate, often siloed systems and under different data structures. These systems include electronic medical records (EMRs), electronic data captures (EDCs), laboratory information management systems (LIMSs), Biobanks, Excel spreadsheets, and legacy databases. Data need to be integrated for every patient across subject areas (demographics, diagnosis, medications, etc.); across studies (clinical trials); and when dealing with omics data sets, across different molecular platforms. The approach to achieve this is to integrate data from the multiple source systems into a single hub, such as a centralized data repository. The extraction of data from a source system and the loading to a target system is undertaken by the extract, transform, and load (ETL) process. Figure 2 depicts sequential steps that are taken in the ETL process as proposed by the eTRIKS consortium [3].

During data integration the following steps are usually taken:

- Master person index (MPI): Typically, patients or subjects may exist in different source systems with distinct identifiers. These need to be resolved to a unique enterprise identifier when loading data into

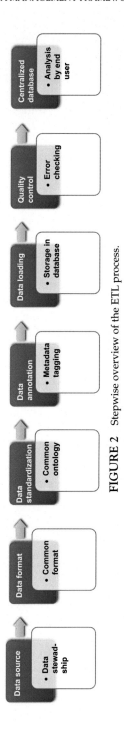

FIGURE 2 Stepwise overview of the ETL process.

a single repository. The MPI will maintain consistency and accuracy when referring to a patient across the enterprise.

- Quality control: Data quality is of utmost importance for accurate, repeatable data analysis, and data mining. When moving data from legacy systems to new ones, it is a good opportunity to clean data as much as possible. Business rules for data assurance and cleansing can be applied in most modern ETL tools. In addition, when collecting data prospectively, a good data quality framework must be in place, which forms part of the organizational data governance.
- Semantic normalization: When centralizing data it is important to normalize semantically to allow sensible querying of data. For example, patient diagnosis data may be coded in different medical terminologies: one clinical source system may use ICD-10 and another SNOMED CT. This can be achieved by providing cross-maps of one terminology to another within the data repository or resolving to the standardized term before loading into the repository.
- Linking omics with phenotype: Besides semantic normalization of clinical data, omics data need to be linked to the phenotype data. Typically this can be achieved by linking sample numbers that reside in Biobank or LIMS systems to patient records, which are in turn linked to research identifiers in omics data sets.
- Cross-platform omics integration: Data generated from multiple omics platforms, such as sequencing machines, and microarray technologies, proteomics platforms need to be integrated into a biologically aware data model. Such a model should be technology agnostic and should be able to accept data using community-based standards, such as the variant call format for loading genomic variants.

Regulatory Compliance and Security

When dealing with health care or life science data, strict regulatory pressures exist. When health care data are used for research, patient confidentiality needs to be protected. The researcher should not be able to identify a patient that has consented to research, so privacy policies must be in place to deidentify (or pseudonymize) patient identifiable information. In the United Kingdom, senior medical personnel referred to as Caldicott Guardians are responsible for protecting patient confidentiality and must ensure that any data management platforms in place meet privacy policies. In the United States, data management platforms that store health care data must comply with HIPAA (Health Insurance Portability and Accountability Act of 1996). Given these regulations, research data and particularly patient-level data must be stored in a secure and compliant manner. Typically there would be a need for certain users with the appropriate privilege to reidentify patients for the purposes of recruitment to clinical trial study. Therefore, the data management framework has to provide that capability.

Role-enabled access control extends beyond patient-identifiable information. Each user group is given controlled access to data as established by collaborators in a governance model. For example, study-level data access should be provided, so that users who have access to study A and not study B be only allowed access to study A data. This level of data access is often applied at the database layer. Another dimension of role-enabled access control could be at the application layer, i.e., the data access user interface. For example, the application may have roles to allow certain users to export data or access certain dashboards. Other types of security could be data encryption on the database layer to prevent unauthorized interceptors from understanding and exploiting the content of the data stored.

Government agencies, such as the Food and Drug Administration (FDA) in the United States and the Medicines and Healthcare products Regulatory Agency (MHRA) in the United Kingdom, require that the industries they regulate, such as pharmaceutical, biotech, and medical device companies, implement controls and safeguards when dealing with electronic data. For instance, it is important to have traceability and audit trails in place when dealing with clinical trial data. In the United States, 21 CFR Part 11 defines the criteria under which electronic records and electronic signatures are considered to be trustworthy, reliable, and generally equivalent to paper records [4]. As a result, any software platform handling electronic data to be submitted to the FDA must comply with 21 CFR Part 11. The European equivalent is Annex 11: Computerised Systems [5].

Scalability

NGS is outpacing Moore's law since 2007 [6], which indicates that the evolution of DNA sequencing technologies is overtaking our ability to generate faster computing architectures. This also leads to an enormous generation of data that need to be stored and managed in scalable platforms. For example, in 2011, the Sequence Read Archive surpassed 100 terabases of open-access NGS reads [7] and the European Bioinformatics Institute (EBI) reported managing 60 petabytes of data, with new biological data doubling every 9 months [8]. And with the announcement of the 100,000 Genomes Project in the United Kingdom in late 2012 [9], it is very evident that scalable platforms will be essential for interpreting whole genome data and for using aggregated data sets for population health and preventive health.

There are multiple ways to achieve scalability in a data management platform. First id by adding more hardware resources to the infrastructure where the data management platform is deployed. Horizontal scaling (scale out) refers to the addition of more nodes to a system, such as a cluster, that are interconnected by high-performance adaptors. Vertical scaling (scale-up) refers to the addition of more resources to a single node in a system, such as more processors to a server computer. Another way to achieve scalability in data management platforms is at the software level. For example, in relational database management systems (RDBMS), such as MySQL (http://www.mysql.com) and Oracle (http://www.oracle.com), partitioning of tables is a widely used technique to achieve scalability when storing large data sets. Furthermore, when the software is deployed in a cluster environment, it needs to have the ability to scale horizontally, such as in Oracle Real Application Clusters that allows multiple nodes to run the database software simultaneously while accessing a single database. With such large data volumes the ability to compress data, while maintaining query power also becomes an important consideration.

DATA MANAGEMENT PLATFORMS

Over the past few years, there have been efforts by the open-source and industry communities to build platforms that will handle the data management and analysis of translational medicine data. Such platforms take advantage of advanced data and knowledge management systems that aim to enable researchers to store, explore, integrate, and analyze omics and clinical data together for generation of new hypotheses or testing existing hypotheses. Several platforms have been developed to support not only the management of a collaborative knowledge through data sharing and provenance but also analysis and visualization. These include BRISK (a package of several Web-based data management tools that provide a cohesive data integration and management platform); cBioPortal (a Web resource for exploring, visualizing, and analyzing multidimensional cancer genomics data); G-DOC (a Web platform that enables basic and clinical research by integrating patient characteristics and clinical outcome data with a variety of high-throughput research data in a unified environment); iCOD (a platform that combines the molecular and clinicopathological information of the patients to provide the holistic understanding of the disease); STRIDE (a research and development project at Stanford University to create a standards-based informatics platform supporting clinical and translational research); TRIAD (a grid-based translational research informatics and data management); Oracle Translational Research Center (a scalable, centralized data storage and analysis solution across genetic information areas, vendor platforms, biological data types, and clinical data sources); and tranSMART (an open-source knowledge management and high-content analysis platform used in translational research for the purposes of hypothesis generation, hypothesis validation, and cohort discovery).

In the area of translational research, several initiatives have undertaken the development of computational infrastructures for creating effective data and knowledge management frameworks. For instance, Informatics for Integrating Biology and the Bedside (i2b2) initiative—funded by National Center for Biomedical Computing (NCBC) in 2004—is a scalable computational framework to address the bottleneck limiting the translation of genomic findings and hypotheses in model systems relevant to human health [10]. i2b2 Hive is a framework composed of software modules that can communicate with each other through Web services and the main modules are responsible for data storage, ontology management, identity management, and others. Although i2b2 Hive is a powerful scalable tool to manage clinical information, it does not support analysis of biomolecular data such as gene expression or nucleotide sequence data. To compensate for these shortcomings, tranSMART and Hadoop frameworks were developed in 2009 and 2011, respectively. tranSMART includes a data warehouse and the associated data mining applications based on open-source systems such as i2b2 and GenePattern [11], whereas Apache Hadoop was developed based on distributed data architecture that Google had used to scale-up from a single server to thousands of machines [12]. This framework has been extensively used by the eMERGE consortium that aimed at high-throughput phenotyping of patient cohorts using EMRs for inclusion in the database of Genotypes and Phenotypes (dbGaP) [13]. The key differences of Hadoop with other frameworks are in computing data through task assignment rather than bringing data into an analytical environment and creating analytical schema from raw data on the fly rather than requiring the ETL preprocessing.

These data management platforms have been adopted by a number of health care and life science organizations. For example, Penn Medicine has deployed a vendor solution that integrates clinical data coming from EMRs with omics data generated from both research groups and diagnostic labs [14]. Pharmaceutical companies and other organizations, including not-for-profits, academic entities, patient advocacy groups, and government stakeholders have adopted the tranSMART platform [15]. As of 2014, there are 32 instances in operation. tranSMART has an extensible data model and an associated data integration framework that allows users to load data. Version 1.1 is built on Grails, a rapid Web application development framework based on a Java platform (http://www.grails.org). This allows data analysis features, such as analytical methods from the R statistical package (www.r-project.org), to be added into the framework. Oracle Health Sciences Translational Research Center (TRC) provides a scalable infrastructure for conducting biomarker discovery and validation [16]. It has been deployed at a number of large academic medical centers worldwide and one of its key features is that it has been developed as enterprise-class software, which means that it leverages core enterprise disciplines, such as security, auditing, archiving, and scalability. A high-level diagram of the architecture used by TRC is shown in Figure 3.

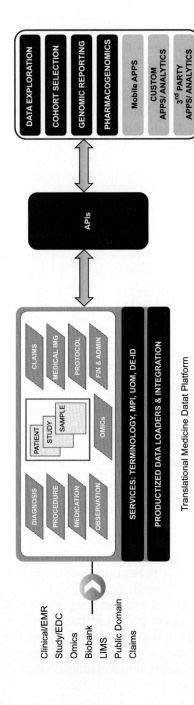

FIGURE 3 The figure shows a high-level diagram of a translational medicine data management and analysis platform. On the left-hand side are the source systems from where data will be extracted, transformed, and loaded into the Translational Medicine Data Platform. The platform integrates clinical, sample, imaging, and omics data and provides terminology, master person index, units of measure, and deidentification services. Then, applications can access the data through an Application programming interface (API). Applications may include data exploration and analysis tools, custom-built applications, and open-source data mining applications, such as R.

DATA MINING IN TRANSLATIONAL MEDICINE

Data mining is a huge topic and so here only a brief introduction will be given to some of the current trends in data mining followed by a discussion on tools and methodologies to get the most out of data mining in the translational medicine setting. The focus will be on which methods can be used to solve specific problems and the technology needed to be successful.

The current interest in data mining is motivated by a common problem across all industries and sectors; how does one store, access, model, and ultimately make sense of very large data sets? The collection of ever-larger stores of data presupposes gaining some insight from collecting all this data and this will generate value. In the case of translational medicine there are several use cases for data mining, some of these are listed below:

- Discovery of pharmacodynamic biomarkers to aid in drug dose determination
- Discovery of patient stratification biomarkers for targeted clinical trials
- Predictive analytics to find patients at risk of complications or adverse events
- Discovery of best combinations for cancer drugs according to the patient's molecular characteristics
- Discovery of causal genes and novel drug targets

A large collection of molecular profiling data with longitudinal clinical data is a fertile ground in which to make such discoveries. As discussed previously, the collection, standardization, cleansing, and organization of such data into large databases is a great starting point for data mining. Many data mining algorithms have the ability to generalize beyond the initial training cohort; this ability is further enhanced when dealing with data that adheres to standards and can be accurately compared. This makes it easier to develop "portable" predictive algorithms, which can then be used with multiple databases and not just on the initial data set they were trained with.

DATA MINING TECHNIQUES AND METHODS

The common goals of data mining applications and methods include the detection, interpretation, and prediction of qualitative and/or quantitative patterns in data. Depending on the nature of data to be mined, quantitative (e.g., gene expression values) or qualitative (e.g., biomedical text), data mining solutions employ a wide variety of techniques ranging from machine learning, artificial intelligence, and statistics to database querying and natural language processing (known as text mining).

For quantitative data mining, there are three classes of algorithms, which can be used.

Supervised Algorithms

These can be used for both classification and prediction, where the set of classes can be labeled and these labels are already known. An example is to predict whether or not a patient will respond to a particular drug given the status or value of some biomarker(s); the two classes here are "responder" and "nonresponder" and the algorithm can use a training set to learn which biomarkers are predictive of each class. Subsequently, this learned model can be used to predict future patients where we do not yet know the outcome.

Unsupervised Algorithms

Algorithms that are used when the class labels are unknown, and we wish to classify a set of patients according to like characteristics; these could be molecular biomarkers or clinical measures. Unsupervised algorithms can determine groupings of the data and thus be used to determine new classes which were previously unknown. An example is taking molecular profiling data from cancer patients and then clustering these patients to find molecular subtypes of breast cancer [17]. Once these new classes are found, we can do further analysis to determine what makes a patient fall into one of these classes and perhaps even predict which class they will be part of.

Semisupervised Algorithms

These methods employ both labeled and unlabeled data to predict the values of missing parameters or extract patterns, clusters, or relationships in the data set. For instance, semisupervised learning has been applied to microarrays to identify cancer biomarkers [18].

Oftentimes data mining methods can be a "black box" where the parameters of the model are meaningless and the optimization of these parameters is purely to achieve better predictive power. Some of the more statistical learning techniques do have some kind of explanatory power and are often used when the data are coming from some kind of known generative distribution, for example, a normal distribution. The art of data mining is to first look at the data and assess whether the data has a complex structure, or is coming from some kind of generative process, which can be modeled using known distributions.

For a thorough theoretical grounding in data mining and statistical learning please refer to Hastie et al. [19] and Hand et al. [20]. In the table below, we list some of the data mining methods used in current translational medicine applications.

There are many other methods than the ones listed in Table 1 and more are being added all the time. If the problem is not tractable by any one

TABLE 1 Machine Learning Methods and Their Applications in Translational Research

Type of method	Exemplars	When to use	Supervised/unsupervised	Examples of use
Decision tree based	CART, boosting, random forest	Classification problems. Random forest is nonparametric and can be applied to many such problems.	Supervised	Classification of clinical outcomes using clinicopathologic variables.
Kernel-based methods	Support vector machines	Useful for 2-class problems and can be extended to N-class problems.	Supervised	Identification of predictive signatures for prognosis using high-dimensional data sets.
Deep learning methods	Convolutional neural networks, deep belief networks	Image recognition or other very complex pattern recognition applications	Supervised	Classification of histopathology images to automate disease diagnosis.
Regression methods	Linear regression, logistic regression, generalized linear models	Good to use when data are coming from a generative process.	Supervised	Building a portable algorithm to detect a certain diagnosis from EMR data.
Clustering methods	K-nearest neighbors, K-means, hierarchical clustering	Exploratory analysis of data to identify groups within the data.	Unsupervised	Clustering of gene expression data to identify molecular subtypes of cancer.
Blind signal separation techniques	Principal component analysis, independent component analysis, nonnegative matrix factorization, single value decomposition	To identify latent factors in the data, which can be used to group items into like groups. SVD is often used to create recommendation systems.	Unsupervised	Exploratory analysis of metabolomics data to visualize groups or clusters. Predicting individual disease risk based on medical history.
Topographic techniques	Self-organizing maps, growing cell structure networks	These methods are useful when you not only want to group the data but also want to visualize the relationship between groups.	Unsupervised	Visualization and interpretation of flow cytometry data.

algorithm alone, it may be possible to use an ensemble of models, which work together. An example is the Heritage Health Prize, which was a competition to build a model to predict which patients would be admitted to hospital and for how many days, based on two previous years of medical record and insurance claims data. The top ranking models were an ensemble of models, which used up to 20 different models that when combined gave the best predictive power [21].

BIG DATA AND DATA MINING

A traditional data mining approach is to extract data from a database, clean it up, and then construct training and test sets for data mining. The learned models are then somehow "productionized" and used to deliver actionable analytics for some kind of application; the model may be updated from time to time with new data to make sure it remains predictive. In the world of big data, it becomes increasingly difficult to move big data sets around; instead the data mining can now often be done in situ where the data never leave the database. Tools such as Oracle Data Miner or Oracle R Enterprise allow this kind of data mining and also enable the resulting predictive models to be persisted in the database; these tools include many of the methods described in Table 1. For unstructured or other kinds of big data (e.g., streaming data from sensors), a data lake architecture could be used (see Figure 4) in which unstructured data are mined with various algorithms using a big data toolset such as Apache Hadoop [12]. Hadoop allows flexible analytics of big data such as text, images, streaming data, and server logs. If any interesting patterns or structures are found, these can be extracted and the structured data can then be loaded into a traditional database for conventional analytics, reporting, and intelligent alerting. Similarly any predictive model could also be operationalized giving the ability to provide actionable intelligence on the unstructured data sets. Data mining algorithms have been rearchitected for the Hadoop system with several platforms now available such as h2o [22], Apache Mahout [23], and Apache Spark [24]. These implementations allow data mining of data sets, which were previously too big to be tackled by traditional systems.

For example, Freeman and colleagues have used an Apache Spark architecture for mapping large-scale brain activity mapping data [25]. Neural activity data can be recorded using techniques such as whole-brain light sheet imaging and two-photon imaging of behaving animals. Such data sets can represent a huge amount of data, and the spark architecture allows exploratory analysis of which methods can be used to represent, reduce, and gain insight from this data. Similarly Apache Mahout and a random forest classifier were used to analyze medical records from

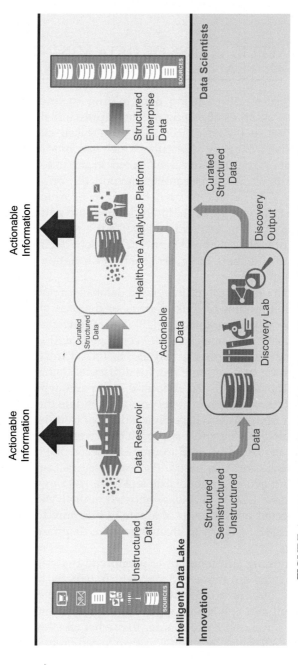

FIGURE 4 A big data architecture that can be used for data mining and production predictive analytics.

patients to detect a diagnosis of amyotrophic lateral sclerosis from a large database of clinical trial records [26].

APPLICATIONS OF DATA MINING IN TRANSLATIONAL MEDICINE

A classic case study of using data mining to find biomarkers is Van't Veer's work on establishing a gene expression signature predictive for survival in breast cancer [27]. Supervised classification was used to establish a 70-gene signature, which can predict good or poor prognosis. This signature was then tested in an independent cohort to determine its ability to generalize beyond the original study [28]. More recently a diagnostic test has been in use, which makes use of this signature as a prognostic tool in order to spare patients from needless chemotherapy. This test and others like it are proving to be a cost-effective way of managing breast cancer treatment decisions, while also giving patients and their caregivers more information in order to try and improve outcomes.

Li and Ruau looked at a large integrated database of published genome-wide association studies to look at pairs of diseases and traits, which shared the same genetic architecture [29]. The rationale for this being that if there is a biological connection between a gene and trait, and a gene and disease, the trait and disease may also be linked and could be used as a way to diagnose or determine risk for a disease. Using statistical techniques, the authors found several disease–trait pairs, which were known already in literature and a few novel ones. To validate the novel findings, they used data from EMRs where clinical data for traits and disease diagnosis were available. Using this approach they were able to validate that some of the traits could be used to predict disease diagnosis sometimes before the disease was diagnosed. For example, mean corpuscular volume (MCV) and acute lymphoblastic leukemia (ALL) were both associated with the gene IKZF1. This was validated in the EMRs and abnormal MCV was detected within 1 year of a patient diagnosed with ALL. Such findings have the possibility to change how we diagnose or detect early signs of disease.

Before performing any genotype to phenotype analysis, it is important to make sure cases are those patients, which actually have the disease in question. In an effort to identify individuals with rheumatoid arthritis (RA), Carroll and colleagues applied a regression model using various parameters from the EMR in order to predict whether a patient truly had RA or not [30]. This algorithm was then tested on two other data sets using different EMR systems to determine portability. The same model with the same parameters was able to predict the RA diagnosis on data from different EMR systems, suggesting that we can achieve predictive models, which can be generalized to other health care systems.

Zhang and colleagues used a novel classification method to group burn patients into different risk groups based on time course gene expression [31]. The use of longitudinal gene expression measurements was more predictive than using any one timepoint on its own. In their method they first compute the *optimal direction* of gene expression and then use standard classifiers to build a predictive model for risk groups.

CONCLUSION

Big data in medicine holds prospect of becoming an engine for the knowledge generation that is necessary to address extensive unmet information needs of patients, clinicians, administrators, researchers, and health policy makers. Particularly in the area of translational medicine where the aim is to convert the findings of bench side into clinical practice in the bedside, integrative data mining frameworks play a crucial role in linking molecule-level data to patient-level information. The benefits of establishing a data management and data mining framework at the organizational level are numerous. It helps the user community, by reducing the time and effort needed to gather data from different departments, integrate them, and preprocess them, so that essential analysis can be performed. At an operational level, it improves efficiency and reduces costs. The provision of audit trails and data access security ensures that the organization complies with regulations, patient confidentiality is protected, and research findings are not lost and can be replicated. Traceability of analytical workflows is also critical to understand which data mining algorithms are useful and where they can be applied. Standards in data management and consistent APIs will lead to portable predictive algorithms, which can be applied in different organizations.

Although several integrative data mining platforms for translational purposes are underdevelopment, there is still a need to develop analytical methods that can simultaneously integrate heterogeneous data types from various modalities and enable efficient interpretations as well as predictions. Once the technological gap between the bench and bedside is addressed with the help of translational bioinformatics tools, including data and knowledge management systems, new horizons for enabling proactive, predictive, preventive, and patient-centerd medicine will emerge. Indeed, effective management and mining of data present ample opportunities for efficient use of information and knowledge in personalized health care. It is foreseen that applying data management and mining strategies particularly to big data can better inform decision making in drug discovery and development during the translation life cycle. As the amount of data and analysis tools proliferates, it will become indispensible for every organization to establish a data management and data mining framework for translational medicine and clinical research. This extends not only to hardware, software, and processes but also to people. Data scientists with domain expertise will become an essential component of this framework.

The proper implementation of the systems described above will also enable new advances at the point of care. The use of precision medicine where genetic and other diagnostic tools are employed to better diagnose and treat patients will gain further prominence. There are some initial signs that these approaches can work and can improve outcomes while minimizing costs. In the United States, Intermountain Healthcare has published the results of a study where the survival time for cancer patients on "genomically guided therapy" nearly doubled that of those receiving standard treatment. At the same time the cost was lower than for standard of care patients [32]. These initial encouraging results have led several governments to put genomic medicine high on the agenda and this has great potential to bring these new advances to those that need them the most: the patients.

LIST OF ACRONYMS AND ABBREVIATIONS

ALL Acute lymphoblastic eukemia
API Application programming interface
EDC Electronic data capture
EMR Electronic medical record
ETL Extract, transform, load
LIMS Laboratory information management system
MCV Mean corpuscular volume
NGS Next-generation sequencing
RA Rheumatoid arthritis
TRC Translational Research Center

References

[1] https://www.amia.org/applications-informatics/translational-bioinformatics.
[2] Total number of retail prescription drugs filled at pharmacies. State health facts. Available at: http://kff.org/other/state-indicator/total-retail-rx-drugs.
[3] http://www.etriks.org.
[4] http://www.fda.gov/regulatoryinformation/guidances/ucm125067.htm.
[5] http://ec.europa.eu/health/files/eudralex/vol-4/annex11_01-2011_en.pdf.
[6] Mardis E. A decade's perspective on DNA sequencing technology. Nature 2011;470: 198–203.
[7] Kodama Y, et al. The sequence read archive: explosive growth of sequencing data. Nucleic Acids Res 2012;40(D1):D54–6.
[8] http://www.ebi.ac.uk/about/background.

[9] http://www.genomicsengland.co.uk.

[10] https://www.i2b2.org.

[11] Scheufele E, et al. tranSMART: an open source knowledge management and high content data analytics platform. AMIA Jt Summits Transl Sci Proc 2014;2014:96–101.

[12] http://hadoop.apache.org.

[13] Chute CG, et al. Some experiences and opportunities for big data in translational research. Genet Med 2013;15(10):802–9.

[14] Wells B, et al. The role of technology in precision medicine. JHIM 2014;28:43–9.

[15] Athey BD, et al. tranSMART: an open source and community-driven informatics and data sharing platform for clinical and translational research. AMIA Jt Summits Transl Sci Proc 2013;2013:6–8.

[16] Ou W, Sheldon J, Oracle Health Sciences Translational Research Center. A translational medicine platform to address the big data challenge. Oracle 2012.

[17] Perou CM, et al. Molecular portraits of human breast tumours. Nature 2000;406:747–52.

[18] Chakraborty D, Maulik U. Identifying cancer biomarkers from microarray data using feature selection and semisupervised learning. IEEE J Transl Eng Health Med 2014;2:1–11.

[19] Hastie T, et al. The elements of statistical learning, data mining, inference, and prediction. Springer Science & Business Media; 2013.

[20] Hand DJ, et al. Principles of data mining. MIT Press; 2001.

[21] http://www.heritagehealthprize.com/c/hhp/details/milestone-winners.

[22] http://h2o.ai.

[23] http://mahout.apache.org.

[24] http://spark.apache.org.

[25] Freeman J, et al. Mapping brain activity at scale with cluster computing. Nat Methods 2014;11:941–50.

[26] Ko KD, et al. Predicting the severity of motor neuron disease progression using electronic health record data with a cloud computing big data approach. IEEE Symp Comput Intell Bioinforma Comput Biol Proc 2014:1–6.

[27] Van 't Veer LJ, et al. Expression profiling predicts outcome in breast cancer. Nature 2002;415:530–6.

[28] Van de Vijver MJ, et al. A gene-expression signature as a predictor of survival in breast cancer. N Engl J Med 2002;347:1999–2009.

[29] Li L, et al. Disease risk factors identified through shared genetic architecture and electronic medical records. Sci Transl Med 2014;6:234ra57.

[30] Carroll RJ, et al. J Am Med Inf Assoc 2012;19:e162–9.

[31] Zhang Y, et al. Classification of patients from time-course gene expression. Biostatistics 2013;14:87–98.

[32] Nadauld L, et al. Precision medicine to improve survival without increasing costs in advanced cancer patients. J Clin Oncol 2015;33:e17641.

Modeling and Simulation Applications in Drug Development Process

Parviz Ghahramani

Chief Executive Officer, Inncelerex, Jersey City, NJ, USA; Affiliate
Professor, School of Pharmacy, University of Maryland, Baltimore, USA
Parviz.Ghahramani@inncelerex.com

OUTLINE

Translational Medicine: Tools and Techniques
http://dx.doi.org/10.1016/B978-0-12-803460-6.00003-9

55

INTRODUCTION

Translational medicine is the link between exploratory research and application of the medicinal products resulting from such a research [1]. Over the last five decades the amount of data and information in the biomedical research has exploded. At the same time the needs and capabilities for computing have exponentially increased.

Modeling and simulation has become a vital approach to summarize and extract information from a huge amount of data in order to provide a tool for exploring new mechanisms of action, diagnosis, and identification of diseases or for prediction and assessment of efficacy or safety of pharmaceutical products. This chapter deals with two major applications of modeling and simulation, one for signal searching in biological systems, which would attempt to identify pathways and to characterize the interplay between them in certain diseases. Another application of modeling and simulation is in drug development from Phase 1–4 to predict or establish the efficacy or safety profile of an investigational drug or to monitor the safety–efficacy profile of an approved drug product.

Figure 1 depicts a schematic view of general drug development phases with emphasis on some examples of application of modeling and simulation that are discussed in more details in this chapter.

THE ROLE OF MODELING AND SIMULATION IN TRANSLATIONAL RESEARCH IN DRUG DEVELOPMENT

Modeling and Simulation in Preclinical Studies (Discovery, IND/CTA-Enabling Studies, and Formulation Development)

Modeling and simulation has made significant contributions in many aspects of preclinical studies in drug development. The application of modeling and simulation starts from early stages of discovery. For example, designing an active molecular structure that would fit a particular receptor with biological activity is a mix of chemistry knowledge and in silico models that would optimize the structure of a chemical entity to achieve desired physicochemical characteristics. Such in silico models enable researchers to design and select a compound with reasonable potential for further development (e.g., good solubility, absorption, target-binding capability). In the next steps of drug development, i.e., animal studies required in the preclinical stage and IND/CTA-enabling studies, modeling and simulation is applied to develop pharmacokinetic (PK) and/or pharmacokinetic–pharmacodynamic (PK-PD) models using the data from animals in order to understand exposure–response relationship

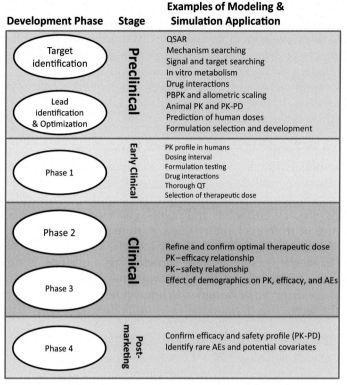

FIGURE 1 Different phases of drug development and examples of modeling and simulation application.

for a drug. Such models will serve as a tool for extrapolation and prediction of optimal therapeutic doses in humans.

Often biomarker data are collected and initial models are constructed using animal data, which are then extrapolated for human predictions. This would provide the first prediction of potential efficacy (or safety issues) related to an investigational drug in an in vivo setting that would also form the basis for optimization and selection of doses for human studies.

There is an enormous amount of information generated in the preclinical stage for a drug development project, this information includes physicochemical characteristics; absorption, distribution, metabolism, and excretion (ADME) properties; time–concentration profile of a drug candidate; in vitro assays; formulation (composition and dissolution); as well as animal PK, PD, and toxicokinetic data. This would provide a collectively complex and extensive set of data. This information is usually completely or partially integrated in models using software packages capable of handling such information to make predictions of human PK, efficacy, and safety profile before any administration of an investigational compound in humans.

Using modeling and simulation, one can even simulate scenarios that are not practical to test in animals or in humans and provide better, faster, and informed decisions in drug development. This approach would reduce the time and cost of drug development and increases likelihood of selecting and developing the best drug candidates for full development.

To illustrate more clearly, this section will provide a few examples of the application of modeling and simulation in preclinical and discovery stage of drug development.

Modeling in Target Identification and Validation

In traditional drug discovery and development the first step is the target identification and validation. Modeling and simulation has contributed in this stage in several forms. Data (such as blood chemistry, physiological effects, efficacy, and safety end points) collected from patients in clinical practice or from exploratory trials in animals or humans are used in models that search for correlations between the variables collected in the data set to identify potential cause-and-effect relationships. In turn, these relationships and their respective pathways may be identified as potential targets that a drug can be designed to interact with to get desired pharmacological effects.

Modeling in Lead Identification and Optimization

An application of modeling for this stage typically includes a molecular modeling based on quantum mechanical principles to design a chemical entity that has a potential to interact with a target (e.g., a receptor). The molecular modeling approaches design a chemical structure that would be an optimal match for the target receptor associated with a desired effect directly or through a cascade of intermediary steps. This approach often is referred to as QSAR (quantitative structure–activity relationship).

Modeling in ADME

During the molecular design stage, a series of molecules with optimal QSAR are selected, and several compounds may be identified as the match for the biological target, but these compounds must be screened further for their ADME characteristics. Approaches used in ADME draw heavily on modeling and simulation methods. An example of the application of modeling and simulation is the use of physiologically based pharmacokinetics (PBPK) to predict PK characteristics of drug candidates. There are many software applications [2] that can provide an initial estimate for ADME characteristics of a compound (e.g., absorption, clearance, volume of distribution, etc.). These applications use physicochemical characteristics of the molecule (e.g., pK_a, lipophilicity, tissue partition, in vitro metabolism data, etc.) as an input into models. Some of these inputs may be informed or validated by established databases of information collected

on other compounds and can fairly accurately predict the initial estimates for PK parameters and PK profile of a compound just by know-
ing the chemical structure. The PBPK models can be further refined and
validated by adding data from animal studies that collect PK data. The
refined PBPK models can be then scaled up (for the size of liver, kidney,
tissue, etc.) to estimate the relevant parameters for human subjects and
to predict human PK profile. The scaling methods are generally referred
to as allometric scaling, which itself is also a type of modeling of the data
for extrapolating from animals to humans. These models could provide a
tool for selection of the most relevant and safe doses of an investigational
compound for use in initial studies in humans. The predictions help to
determine the lowest efficacious dose and top therapeutic dose required
to be investigated in human subject studies to characterize efficacy–safety
profile of the investigational compound.

Use of Modeling and Simulation in Phase 1

The doses of an investigational drug used in first human studies are
typically determined based on the PK information from animal studies
(see previous section). The first dose in a first human study is normally
selected to be below therapeutic threshold and is administered to a cohort
of subjects followed by escalating doses sequentially administered to separate cohorts of subjects. Normally, PK and safety data are collected and
assessed after each dose level and before escalating to the next dose cohort.
In the process of this assessment, PK or PK-PD modeling is applied often
to refine the predictions for the amount of exposure or pharmacological
effects expected in the next dose cohort of subjects, and adjust the dose if
necessary. At the end of Phase 1, a better understanding of the PK profile
in humans emerges with the help of modeling and simulation that will
be extrapolated to select doses with the best therapeutic potential for a
desired efficacy and the least side effects in Phase 2.

Data from early PK studies are often used in PBPK or conventional
compartmental models. These models play an important role in determining time interval between doses of the compound to achieve optimal
therapeutic effects. For example, collection of PK data and modeling the
information at preclinical stage would allow prediction of the half-life of
a drug candidate for oral administrations in humans. The half-life estimate in turn would determine if adequate exposure over time may be
achieved after a twice daily administration. If the twice daily administration shows inadequate exposure, one may be able to develop and select
an extended release formulation that would contain a higher total daily
dose that would be released slower over a longer period of time. This may
provide an approach to achieve a twice daily dosing regimen that would
cover the exposure to the drug adequately between doses.

Models of PK, PK efficacy, PK side effect, or PK biomarker that have been developed in preclinical studies are updated and refined by collecting data in Phase 1. The models and the predictions would characterize PK (and sometimes PK-PD) profile of the investigational drug in human subjects (most often healthy volunteers). This information is utilized to predict and select the most relevant therapeutic doses for further testing in Phase 2 (normally the intended patient population). For some compounds, it is possible to get a first glance at the potential efficacy (or side effect) profile of the drug by measuring biomarkers and their relationship with drug exposure even in Phase 1. Subjects in Phase 1 are most often not the target patient population but can provide biologically relevant effects or markers that are linked to efficacy in patients. The models developed at this stage integrate and take into account all the previous information in the preclinical phase as well as Phase 1 human data.

Biomarker analysis in Phase 1 is particularly useful because the underlying mechanism is tested for the first time in human subjects with no extrapolation from animal data. Discontinuation of development projects in Phase 1 most often is due to major safety or tolerability findings. However, in some cases lack of expected effects (for example, changes in an established biomarker), that is indicative of clinical efficacy, could result in termination of a program even in Phase 1. An example is the development of pegylated compounds such as epoetin. It is known that erythropoietin family of compounds, following a few weeks of administration, would cause an increase in red blood cell count even in healthy volunteers. It is, therefore, expected that any compound with similar mechanism of action (e.g., pegylated epoetin) would also induce some increase in red blood cell counts in healthy volunteer subjects in a Phase 1 study. Therefore, decision to continue (or discontinue) the development of a pegylated epoetin could be made after collecting PK and red blood cell data from merely Phase 1 studies. In this case red blood cell count is an obvious biomarker, which has a direct correlation with the clinical end point measured in the target patients. In other cases, a biomarker might be more indirect (i.e., distant from the cascade of events leading to a therapeutic event), but may still provide an early insight into the likelihood of its efficacy or side effects. The value and the reliability of such biomarkers are determined by the amount of historical data and validation available for correlation between the biomarker and the desired clinical efficacy end point.

Another area that modeling and simulation has played a substantial role is drug–drug interactions. Most often potential drug interactions are explored and determined using in vitro data that would inform drug interaction models to predict the probability and the magnitude of an interaction. The initial data are most often collected in the form of in vitro metabolism (e.g., metabolism by CYP450s) or in vitro inhibition studies. Predictions of drug–drug interactions have resulted in recommendations in drug labels that are in some cases purely supported by modeling and simulation without conducting a drug–drug interaction study in human volunteers [3]. Given the large number of potential drug–drug interactions, especially for drugs metabolized via CYP450s or for drugs that affect transporters, it is impractical and costly to perform numerous drug interaction studies in humans. Modeling and simulation has provided a tool to minimize the number of drug interaction studies needed to be performed in humans and has had a significant impact on reducing time and cost of development in this respect. Modeling and simulation allows collective knowledge of in vitro metabolism studies, combined with a limited number of drug–drug interaction studies in humans, to predict a potential interaction, and its magnitude, with many other drugs that may never be tested directly in human drug interaction studies.

Another area benefited significantly from modeling and simulation is the investigation of drug effect on cardiac repolarization, which is concerned with drugs that may cause QT prolongation. QT prolongation is commonly associated with *torsade de point*. QT prolongation could be a single factor leading to discontinuation of a drug under development or withdrawal of a drug that is already marketed [4]. Modeling and simulation has provided a range of approaches to characterize drug effects on QT prolongation. Early signals of QT prolongation can be detected by modeling the drug concentration-QT data collected during routine Phase 1 studies. In cases where the modeling of the exposure-QT data would detect a signal, it can also provide information about the maximum potential prolongation at different doses and whether the observed prolongation is clinically relevant. Modeling and simulation has been used to provide guidance for QT prolongation associated with a given dose of a drug, which is actually not investigated per se in a QT study in humans. An example of such a case is when QT prolongation is predicted for a dose, which is not studied directly in a human trial, using an established relationship between drug concentration and QT prolongation from existing human data [5].

Modeling and Simulation in Phases 2, 3, and 4

Data collected in Phase 2 and 3 studies are mostly from patients. PK and clinical data (efficacy and side effects) from these studies would provide a good understanding about the relationship between exposure and effects of an investigational drug. The main aim of Phase 2 studies is to find the optimal therapeutic dose (the so-called dose-ranging studies) and then to confirm the efficacy and safety profile of the selected dose(s) in Phase 3. Modeling and simulation has been applied successfully to play a pivotal role in the selection of the right dose before going into Phase 3 studies. This is normally done by modeling Phase 2 data and integrating

Phase 1 data (and may be even integration preclinical data). Modeling can help to reduce the number of doses that must be investigated in Phase 2 studies or at least help to select the most likely doses that are therapeutically viable. The results of such modeling could enable to remove one dose (one arm) in a study significantly reducing the cost and time needed for a Phase 2 study. Similarly, modeling and simulation would help to provide definitive confirmation for the therapeutic doses in Phase 3.

Modeling would provide a powerful tool to analyze integrated data from Phase 1–3 to identify subpopulations that may respond differently in terms of their clinical efficacy or side effects. Modeling would provide a more comprehensive understanding of the effect of demographic factors (such as age, gender weight, geographical region) on the exposure, clinical efficacy, and side effect profile of the drug.

In Phase 4 (postmarketing studies), modeling and simulation has been traditionally not utilized in its full potential. But, modeling has been used to provide further confirmation and refinement of therapeutic doses and to identify subpopulations. However, commercial incentive, to refine and find new doses for therapeutic use of a drug that is already approved, has been limited and, therefore, this area has had less application of modeling and simulation compared to Phases 1–3 in general.

Impact of Modeling and Simulation on Drug Approvals

During the last two decades modeling and simulation has been increasingly used in drug applications submitted to regulatory authorities and has made an impact on approvals and drug labels [6]. However, a higher level of impact is in progress where modeling and simulation is playing a role in approval of drugs and labels based on modeling of biomarker data in lieu of clinical outcome. This approach is specially being established for indications where the real clinical end point would take several years (4–10 years) to measure in a traditional drug development program. For example, for a new Alzheimer's drug development program to result in a commercially successful product, the drug is required to show clinical efficacy with a disease-modifying effect. Traditionally, the efficacy end point in clinical development studies in Alzheimer indication has been required to show a significant slowdown in the rate of cognitive decline in Alzheimer patients. Such studies are expensive and would take several years (4–10 years) to complete. Moreover, normally the sponsor would not get any early indication if a drug is likely to fail the clinical efficacy, or if the drug would meet the requirement for approval by the regulatory authorities, until the end of a full clinical program. This would mean a massive investment with a relatively unpredictable outcome. This has deterred many sponsors to continue research in such therapeutic areas and, therefore, drug development programs for diseases like Alzheimer are in decline in the recent decades.

This is in the face of the unmet medical need for a large population of patients who suffer from debilitating diseases such as Alzheimer.

Recently the US Food and Drug Administration initiated a public–private program [7] that has employed modeling and simulation to a large extend to analyze data across clinical studies and to understand disease progression, the time course of disease, covariates, and the biomarkers related to the diseases. Alzheimer is one of the cases chosen for this initiative. The aim of the project is to identify early biomarkers that could be of predictive value for the long-term clinical efficacy end point in Alzheimer. A success in this approach would mean sponsors can expect to invest in Alzheimer products with a reasonably lower risk of failure, i.e., being able to utilize biomarker data to assess the likelihood of success for getting drug approvals earlier in the clinical programs. This would mean a potential reduction in the duration and the cost of clinical studies. Obviously, the regulatory authorities may still require that the sponsors do longer term follow-up trials post approval, but the approval could be mainly based on positive results of changes in biomarkers rather than long-term clinical efficacy.

Oncology, a Special Case

Oncology deserves a special attention in this chapter due to its advances in some particular aspects in the application of modeling and simulation. In the last few decades, research has identified numerous mechanisms for cancer etiology and has determined many different types and subtypes of cancer. The scientific knowledge in this field has grown enormously. With this growth, comes the explosion of volume of information generated on biomarkers, genetic markers, pathways, and interaction between these, which in turn has created a large amount of data now being used in modeling and simulations. The modeling and simulations are done for different purposes. These include signal searching to correlate a particular type of cancer with specific biomarkers, demographic factors, or other characteristics. The modeling and simulation also includes PK and PK-PD analysis and assessment of correlations to optimize the treatment, similar to the approaches described in the earlier section in this chapter.

Given the seriousness of the impact of cancer on patients' life, there is a great deal of tolerance for side effects and even for toxicity in cancer treatments. This means any treatment with reasonable efficacy in cancer treatment could be considered a successful drug regardless of side effect profile (as long as the side effects are manageable and not life threatening). Modeling and simulation has been successfully applied to establish maximum exposure and to provide a tool to maximize the number of patients who respond to an anticancer treatment. However, the research and the number of centers that work in the area of cancer research continue to increase. Given the large variety of cancer types and numerous mechanisms leading to cancers, the amount of data available in this field is increasing exponentially every year. It is ultimately more appropriate to use modeling and simulation in oncology to identify relevant signals and mechanisms that cause cancer. But, also it is equally important to apply more PK and PK-PD modeling to get maximum efficacy from existing drugs that have been approved for marketing already as well as those that are in development for cancer. The principles of application of modeling and simulation in oncology are not too dissimilar to that described earlier in this chapter. But, the main difference is more emphasis on efficacy with less focus on side effects and tolerability.

NEXT LEVEL OF CONTRIBUTION FROM MODELING AND SIMULATION

Until recently, modeling and simulation has been effectively used to support decision making at the study level and at project level. At study level, modeling has been used to design more efficient studies, for optimal dose selection and in the interpretation of study results and in assessing the interplay between various patient characteristics (demographic factors, geographic region, ethnicity, etc.) and clinical outcomes. At the project level, modeling and simulation has had impacts on go/ no-go strategic decisions for a project and on drug approvals and drug labels. However, the next generation of impact (which is not yet fully in use) could be the application of a systematic modeling and simulation in portfolio management. Although some quantitative approaches have been employed in portfolio analysis and prioritization, these are normally done rather in isolation from the modeling and simulation that is routinely done at the study and project levels. A higher level of efficiency could come from portfolio analysis models that get their primary quantitative assumptions and inputs (probability of success, variability, etc.) directly from the output of the modeling and simulation done at the study and project levels throughout the drug development [8]. This type of application of modeling and simulation is still in its infancy (if at all in use) at present. This approach has the potential for a large impact on organizations that need to decide and prioritize among several potential projects in their portfolio. Such an approach needs an integrated modeling and simulation that include input from marketing, business development, as well as research and development functional inputs. Figure 2 shows different levels of contribution and application of modeling and simulation at different levels. While drug development organizations have utilized a great deal of modeling and simulation at the study and program levels to optimize the outcomes and decision making, the application of an integrated modeling and simulation in portfolio management is yet primitive. The portfolio management

FIGURE 2 Applications of modeling and simulations at different levels in drug development.

has little benefit in smaller organizations with limited number of projects, but has a great potential value for larger drug development organizations where there are choices to be made among several projects competing for limited resources.

References

[1] Cohrs RJ, Martin T, Ghahramani P, Bidaut L, Higgins PJ, Shahzad A. Translational medicine definition by the European Society for Translational Medicine. New Horizons Transl Med 2014;2(3):86–8.
[2] Poirier A, Funk C, Lave T. Role of PBPK modeling in drug discovery: opportunities and limitations. Proc Appl Math Mech 2007;7:1121907–8.
[3] Zhang L, Zhang YD, Zhao P, Huang SM. Predicting drug-drug interactions: an FDA perspective. Am Assoc Pharm Sci J 2009;11(2):300–6.
[4] Roden DM. Drug-induced prolongation of the QT interval. N Engl J Med 2004;350:1013–22.
[5] Khariton T, et al. Population exposure-response analysis for an escitalopram thorough QT study. J Pharmacokinet Pharmacodyn 2013:S64–5.
[6] Lee JY, Garnett CE, Gobburu JV, Bhattaram VA, Brar S, Earp JC, et al. Impact of pharmacometric analyses on new drug approval and labelling decisions: a review of 198 submissions between 2000 and 2008. Clin Pharmacokinet 2011;50(10):627–35.
[7] FDA website. CAMD initiative, http://www.fda.gov/AboutFDA/PartnershipsCollaborations/PublicPrivatePartnershipProgram/ ucm231134.htm [accessed 09.05.15].
[8] Ghahramani P. Modelling and simulation the conduit connecting translational medicine with portfolio management. Oral presentation. In: European Society for Translational Medicine (EUSTM), 2014. September 22–25, Vienna. 2015.

CHAPTER

4

Advancements in Omics Sciences

Adriana Amaro[1], Andrea Petretto[2], Giovanna Angelini[1], Ulrich Pfeffer[1]

[1]Molecular Pathology, IRCCS AOU San Martino–IST Istituto Nazionale per la Ricerca sul Cancro, Genova, Italy; [2]Core Facility, Istituto G. Gaslini, Genova, Italy

OUTLINE

Translational Medicine: Tools and Techniques
http://dx.doi.org/10.1016/B978-0-12-803460-6.00004-0

67

OUTLINE OF THE CHAPTER

In this chapter we try to give an overview of the recently developed omics technologies in a tentatively exhaustive fashion. Each aspect of the technologies will be presented in a manner understandable for the student or the professional with a basal knowledge of modern biology and medicine. The relevance of the technologies for translational medicine will be described recurring to relevant examples. The reader will be enabled to understand reports and to propose scientific approaches using these technologies. The limitations and current pitfalls of the approaches will be critically discussed.

INTRODUCTION

The knowledge of the complete sequence of the human genome and the presence of an ever-growing amount of biological data present in many data repositories allow for analytical approaches that consider the entirety of biological domains instead of single items. The accumulation of information is accompanied by the development of machines, such as microarrays, liquid handlers, lab-on-a-chip, and microfluorimeters, that are able to parallelize and automate the analysis of biological molecules and their interactions as well as by the development of bioinformatic approaches to mine the huge amount of information generated by high-throughput analyses. Biology, engineering, and bioinformatics have thus joined in a virtuous synergy that has generated the "omics" approaches.

On the one hand, this has conceptual consequences in terms of investigative approaches that were heretofore driven by a hypothesis generated by the scientist on the base of existing knowledge and are now driven by the data themselves or the domain knowledge, knowledge contained in biological repositories, and large-scale data collections not necessarily known to a physical person. The shift from hypothesis-driven to data-driven research is of particular importance in biology where falsification of scientific hypotheses as postulated by Carl Popper is hindered by the complexity and diversification of biological systems that rarely allow for stable generalizations.

Omics approaches avoid a priori reduction of biological complexity, although up to the present, downstream analyses often reintroduce reductionism given the fact that systems biology lags behind: real consideration of complexity remains in the future of biology. In translational medicine, omics introduces three new aspects: (1) the ease of parallelization allowing for the analysis of many if not all biological molecules potentially involved in a given pathological state, (2) the possibility to identify pathological drivers among a large variety of potential actors, and (3) the personalization of medicine that takes into account the precise individual condition including germ line and somatic genetics as well as features acquired during lifetime.

The physician of the future will ever more rely on molecular information in terms of diagnostic, prognostic, and predictive classifiers that will guide his/her clinical decision in complementation of her clinical experience. Omics generates the promise of targeted approaches from disease prevention over the treatment of overt disease to the management of chronic conditions and resistance to therapy. In the recent future, no physician will be able to exert his/her profession without a structured understanding of the omics technologies. This chapter will therefore introduce the reader to the main areas of omics through a brief description of the technologies and their application in translational medicine.

The main omics domains are genomics, epigenomics, transcriptomics, proteomics, and metabolomics. Figure 1 shows the information flow from

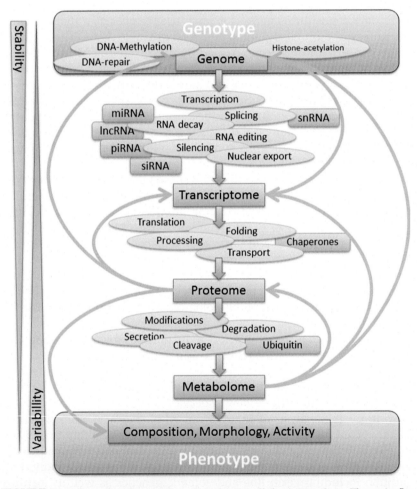

FIGURE 1   **Information flow in the cell and intracellular interactions.** The main flow of information in the cell is unidirectional leading from the genotype to the phenotype. Nuclear DNA is exposed to modifications, mainly DNA methylation, and the availability of the information is also influenced by modifications of histones bound to the DNA. The

the genome to the phenotype together with a selection of important processes involved and some additional molecular players. The information flow corresponds to gradients of stability (genome > transcriptome > proteome > metabolome) and variability (in opposite direction).

GENOMICS

Genomics is the most evolved Omic technology due to the very nature of nucleic acids that are distinguished by the sequence of nucleotides of which they are composed of but not by highly variable chemical and physical properties making the development of analytic techniques straightforward. Almost all techniques for the analysis of nucleic acids are based on A–T and C–G base pairing that allows for the formation of double strands with complementary sequences, one of which can be a probe generated for analysis with the other strand being the sample to be analyzed. The second basic feature of nucleic acids techniques is the use of biological enzymes, mainly polymerases that can synthesize a nucleic acid strand by copying a leading strand thereby revealing its sequence or simply amplify a given DNA molecule in order to reach an amount that can be measured.

Nucleic acid-based omics are used to analyze DNA and RNA, the genotype, and its expression into protein encoding mRNAs or regulatory noncoding RNAs (ncRNAs) of various types.

◀ whole genome is under constant control of DNA repair systems that can repair strand breaks and base substitutions. The genetic information encoded by the genome is transcribed into messenger RNA (mRNA) and other noncoding RNAs (ncRNAs). Transcribed RNAs are spliced, edited, and exported from the nucleus and their stability is influenced by ncRNAs. The resulting group of mRNAs constitutes the transcriptome. Translation of the mRNAs leads to polypeptides that are folded, processed, and transported to their cellular destination with the help of chaperones and constitute the proteome. Proteins undergo many types of posttranslational modifications, are cleaved by proteases for activation and secretion, and are eventually directed to degradation after ubiquitination. Among the mature proteins there are a large number of enzymes that catalyze biochemical reactions in the cell leading to a variety of metabolites such as sugars, lipids, amino acids, and nucleotides that constitute the metabolome. All these processes together determine the cellular composition, morphology, and activity that together define the cellular phenotype. The various "omes" are interrelated through feedback mechanisms and interdependence. These processes are dependent on the information stored in the genome and are mainly executed by proteins and RNA–protein complexes. The direction of the flow corresponds to decreasing stability and increasing variability. The genome is stable in the lifetime of an organism and is essentially identical in all cells of the organism. The transcriptome is made of some tens of thousands of mRNAs and a still not reliably measurable amount of ncRNA species. The transcriptome is relatively stable: it responds to a variety of signals but maintains a profile that is characteristic for the cell. The proteome introduces an additional level of variability and can, especially through posttranslational modifications, rapidly respond to any type of challenge. The metabolome is a highly sensitive and rapidly changing collection of small molecules produced by the cell in accordance with the physiological state of the cell and its environment.

Genomic Technologies

Microarrays

Microarrays are miniaturized arrays of biological probes deposited on a solid surface such as glass or silicon where specific probes of known sequence and reactivity are located in predetermined positions. Initially, microarrays were produced by spotting nucleic acids directly onto reactive surfaces; today, in situ synthesis of oligonucleotides by photolithography or ink-jet delivery of nucleotides in controlled synthesis iterations is commonly performed.

In first-generation microarrays, cDNA clone collections obtained during the human genome project were used for spotting, which were later substituted by ad hoc-produced polymerase chain reaction (PCR) amplicons. The former have a clear limit in the discrimination of different members of gene families with more or less extended regions of homology, whereas the latter present limitations due to the process of PCR and the subsequent need for purification. At present almost all microarrays use synthetic oligonucleotides as probes that can be designed to guarantee specificity and hybridization performance.

The sample cDNA or cRNA is labeled through the incorporation of modified nucleotides, most often cytosine, that either contain fluorophores (direct labeling) or, more commonly, biotin residues to which (strept-)avidin-coupled fluorophores can bind in a separate labeling reaction. The preparation of sample DNA and RNA contains amplification steps needed to generate sufficient signal from reduced amounts of starting material.

Labeled cDNA or cRNA is hybridized to the microarray under ionic strength and temperature that guarantee efficient and specific hybridization followed by washing steps in order to eliminate unbound or nonspecifically bound sample. After washing, the arrays are read in a fluorescence scanner that reveals binding of sample cDNA or cRNA to the hybridization probes on the solid surface in a quantitative manner.

After application of a grid that identifies the single probes in the corresponding localizations on the array, the signal for each position is integrated. Final intensities in terms of copy number of DNA or RNA expression level are obtained after background subtraction, normalization, and eventually summarization of signals from different probes specific for the same gene or transcript. The intensities can then be compared for arrays derived from different samples that are normally analyzed using several biological replicates in order to control for variability within a condition. The analysis pipelines contain software with statistical algorithms dedicated to the different applications of microarrays. For a more detailed introduction into microarray technology, see http://grf.lshtm.ac.uk/microarrayoverview.htm.

Quantitative PCR

Quantitative polymerase chain reaction (qPCR) is based on real-time monitoring of the amplification process as it occurs. This is obtained by using either intercalating dyes such as SYBR Green or probes containing fluorophores in addition to classical PCR primers, the template, the nucleotide-containing reaction mix, and the thermostable polymerase. Many different types of qPCR probes exist that are not detailed here.

Monitoring during the amplification reaction allows for obtaining amplification signals during the exponential phase of amplification in which the material doubles (or nearly doubles depending on the efficiency of the reaction) in each cycle before the reaction components are exhausted. Instead of measuring the amount of amplicons produced at the end of the reaction (end point PCR), the number of cycles needed to reach a predefined level of fluorescence (threshold cycle, ct) is measured. The later a sample reaches the ct, the less specific template was present in the sample. For DNA copy number measurements, the ct of the sample DNA (i.e., the genomic region to be analyzed) is compared to the ct of presumed "normal," diploid genomic regions. For mRNA expression analysis, the ct of the sample cDNA (obtained after inverse transcription of the sample RNA) is compared to the ct of the so-called housekeeping genes for which stable expression across all experimental conditions is assumed. This allows for a relative quantification of mRNA expression in pairs or series of samples. Absolute quantification can be obtained through the application of calibration curves obtained by the amplification of known amounts of template.

A specific application of qPCR for the determination of chromosome copy numbers with high resolution is multiplex ligation-dependent PCR amplification (MLPA) [1]. This method relies on the use of synthetic probes which have a combination of a gene-specific part with an artificial sequence. Two such probes are designed so that their gene-specific parts are adjacent to each other. Upon ligation a single DNA molecule is formed that can be amplified using primers complementary to the artificial sequences incorporated in the probes. Due to the fact that amplification relies on prior ligation of correctly positioned fragments, unspecific reactions are mostly suppressed. This allows for multiplexing in order to analyze several diagnostic cytogenetic events in a single reaction with elevated sensitivity and specificity.

For further information on qPCR, see http://qpcr.gene-quantification. info/.

Next-Generation Sequencing

Before 1977, DNA sequencing was very inefficient and unreliable. In 1977, Maxam & Gilbert and Sanger separately introduced two different sequencing methods, chemical sequencing and dideoxy chain termination sequencing, respectively. Sanger sequencing has become the gold standard for sequencing, and the entire human genome sequencing project has relied on this technique. In next-generation sequencing (NGS), we define sequencing methods that allow for obtaining sequence information for single DNA molecules in a highly parallelized manner [2].

Sequencing of single DNA molecules has been achieved through the inclusion of single DNA molecules bound to beads together with all reaction compounds in droplets of an emulsion. The single DNA molecules are then amplified by a classical PCR reaction, the emulsion is broken, and the beads now containing many molecules derived from one starting molecule are deposited in a microwell plate each well of which can accommodate a single bead. For the most diffused NGS systems, sequencing reactions occur in an iterative way in these wells after addition of the single nucleotides that are incorporated by the polymerase-releasing pyrophosphate that can be measured in a light-producing reaction (454 method) or through the corresponding change in conductivity of the reaction fluid (Ion Torrent). An alternative way consists in binding of the template molecules to a solid surface followed by in situ amplification and polymerase-based sequencing with the introduction of the four nucleotides coupled to different fluorophores (Illumina) (Figure 2). The progress of sequencing is monitored by high-resolution cameras that observe incorporation of nucleotides for each physical position or well on the array and store the sequences in data files for further analysis. Hundreds of thousands to tens of millions of sequences can be read in parallel.

Downstream analyses must filter the sequences according to quality criteria. Given a certain level of misincorporation of nucleotides, a threshold of the number of times a specific sequence must be read, in order to be considered reliable, must be defined. Particularly problematic are mononucleotide repeats. After filtering, the sequences are aligned to the reference genome and the single reads are attributed to specific genes or transcripts and eventual mutations are reported with their relative frequency.

Due to persisting quality issues, NGS sequences still require confirmation by Sanger sequencing. Third-generation sequencing approaches should soon become available. They are expected to determine a further reduction in sequencing costs and will hopefully overcome quality issues related to NGS.

For further information, see https://www.ebi.ac.uk/training/online/course/ebi-next-generation-sequencing-practical-course/what-you-will-learn/what-next-generation-dna-.

Array-Based Comparative Genome Hybridization

Array-based comparative genome hybridization (aCGH) allows for the assessment of the DNA copy number present in the cells of a tissue. For

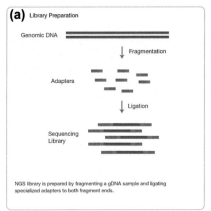

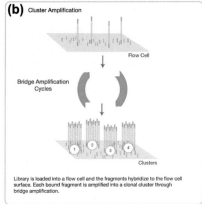

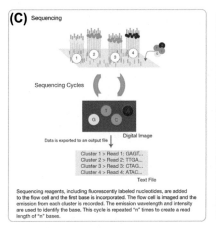

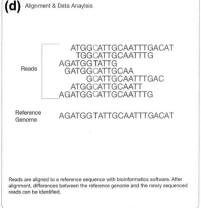

FIGURE 2 **Workflow in next-generation sequencing (courtesy of Illumina Inc.).** The workflow is composed of library preparation (a), cluster amplification (b), the sequencing itself (c), and data analysis (d).

this purpose, signals obtained after hybridization of sample DNA fragments to cDNA or oligonucleotide probes on the array are compared to the signals obtained from a collection of standard DNA. This approach is limited by regions containing duplicated or deleted DNA stretches even in "normal" DNA, a fact that is more frequent than previously assumed. This limit is overcome by the use of larger collections of DNA obtained from healthy subjects of different populations [3] (Figure 3).

The resolution of aCGH arrays depends on the number and the distribution in the genome of the probes. The number of probes also affects the reliability of the analysis, since most analysis methods define a window containing a variable number of probes in order to correct for probes that yield outlier signals.

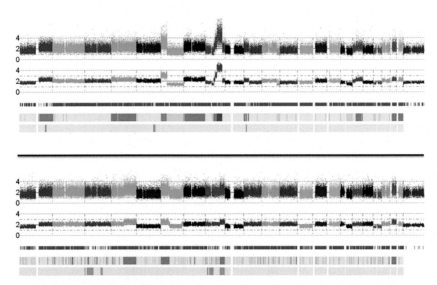

FIGURE 3   **Array-based comparative genome hybridization (aCGH).** The figure shows two typical profiles of an aCGH analysis on human tumor samples. The 23 chromosomes are deployed from the left to the right. The upper dot diagram shows actual data for each single SNP analyzed. The same analysis is shown in the lower dot diagram but in this case, the SNPs are collapsed to a window each containing a user-defined number of SNPs thereby reducing noise. In these two diagrams the copy number is indicated on the left (2 = normal diploid content). The green bar diagram shows position and number of the probes. The lower bars show interpretations of the signals observed according to a threshold defined by the user. The upper bar shows normal diploid regions in yellow, amplified regions in pink, highly amplified regions in red, and deleted regions in light blue. The lower interpretation bar indicates zygosity: yellow, heterozygous; light blue, isodisomy (uniparental disomy) where one copy is deleted and the other copy is duplicated. Amplification of the short arm of chromosome 6 and the long arm of chromosome 8 are visible. The upper case shows additional amplifications and deletions, and the lower case shows many short amplifications and three extended regions of isodisomy. Data from two cases of uveal melanoma obtained through Affymetrix 250k arrays, analysis with copy number analyzer for Affymetrix GeneChip (CNAG) [92].

aCGH arrays allow for the identification of duplication or amplification as well as deletion events. Translocations and gene inversions cannot be revealed. If the probes are designed for the distinction of allelic variants (single nucleotide polymorphisms, SNP; see below), events of gene conversion, i.e., the loss of one allele and the duplication of the other leading to uniparental disomy or isodisomy, can also be revealed.

Genome-Wide Association Studies

Genome-wide association studies (GWAS) can be performed using arrays containing hybridization probes specific for allelic variants (SNPs) occurring in the population. In the most frequent case of biallelic variants, where in a given position of the genome two different nucleotides can be present in a given population with characteristic frequencies, a heterozygous sample yields a signal on both probes, whereas homozygous samples yield a (more intense) signal on only one probe. The genotype for many SNPs can be defined in this way with considerable reliability. Present SNP arrays carry probes for more than a million SNPs. Probes are selected for marker SNPs located in genomic stretches (haploblocks) with a very limited recombination frequency so that the genotype of other SNPs within the same haploblock can be imputed [4].

For further information see, http://www.nchpeg.org/bssr/index.php?option=com_k2&view=item&id=69:genome-wide-association-studies-gwas&Itemid=119.

Innovative Sequencing Approaches

NGS can be used for a variety of sequencing projects with different aims [5]:

- *Whole-genome sequencing* reveals the entire genomic sequence of the genomic DNA obtained from a biological sample.
- *Exome sequencing* [6] is restricted to the sequences of the protein-coding parts of the genome, the exons, which are captured by binding to capturing probes in solution or on arrays prior to the sequencing reaction. Since genomic fragments are captured, normally some information on the regions adjacent to the exons is also obtained.
- *Deep sequencing* [7] exploits the fact that single molecules are sequenced. Therefore, it is possible to obtain the sequences of mutations that are present only in part of the cells as it occurs in the case of mosaicism or somatic mutations (tumor heterogeneity). The depth of sequencing is determined by the number of sequences obtained for each genomic fragment (coverage) and the intrinsic error rate. At present, mutations present in at least 0.1% of the cells can be reliably identified.
- *Targeted resequencing* [7] consists in the (deep) sequencing of selected genomic regions such as those known to contain hotspot mutations or frequently mutated genes.
- *Paired end sequencing* allows for the analysis of long DNA stretches from which partial sequence information is obtained. The alignment of these sequence elements to the reference genome can identify sequences in noncompatible positions as they occur after translocation, inversion, or large insertions or deletions.
- *RNA-sequencing* (RNA-seq) [8] is the application of NGS to cDNA obtained from cellular RNA preparations containing mRNA or all kinds of ncRNAs (lncRNA, miRNA, piRNA, snRNA, snoRNA).

RNA-seq reveals mutations in the coding regions and can be used for the quantification of transcripts, since the number of RNA molecules present in a sample and the number of sequence reads obtained are correlated to a good approximation. Since RNA-seq is not guided by predefined probes, it can detect nonanticipated events of alternative splicing or RNA editing as well as fusion transcripts formed after chromosome translocation and trans-splicing events.

- *Chromatin immunoprecipitation sequencing* (ChIP-seq) [9] is a research method used to identify genomic regions to which specific proteins such as transcription factors bind. Instead of isolating naked DNA, chromatin fragments (nucleosomes) are isolated and reacted with the protein (transcription factor) of interest and specific antibodies against this protein. The complexes formed are precipitated by binding to anti-antibodies coupled to agarose beads; the DNA is isolated and sequenced. The sequence reads obtained correspond to the genomic regions specifically recognized by the DNA-binding protein and can identify gene sets regulated by a specific factor.
- *Metagenomics* [10] is an NGS applied to complex biological samples containing several or many different organisms (mostly micro-organisms) as they are found in environmental samples or stool probes. In order to obtain the list of organisms present in the sample as well as their relative abundance, either a genomic region of high variability flanked by well-conserved regions is amplified (such as 16S ribosomal RNA fragments) or simply all DNA fragments present in the samples are amplified. In the second case, alignment of the sequences with all potential reference organisms is necessary.

Applications of Genomics in Translational Medicine

Detection of Mutations and Variants

NGS is widely applied in translational medicine since most diseases are either caused or influenced by the genetics of the organism, by somatic mutations, or by epigenetic alterations of DNA (DNA methylation). NGS therefore is a major passage in any translational research project [11].

Applications of NGS in translational medicine include the identification of new mutations for congenital diseases without known causative mutations or in the absence of such mutations in cases with known disease genes. Early clinical application of NGS for congenital diseases has successfully been performed for cases without known diagnostic mutations and a phenotypic presentation similar to cases that carry known mutations for which a treatment is available. In several cases, this approach has led to the identification of new mutations in the same pathway of the known mutation thus indicating a therapy normally restricted to carriers of the known mutation (see for example [12]). Where such treatments are not available, NGS remains feasible in a research setting. In this case, and especially in cases where the clinical presentation does not resemble phenotypes with known mutations, sequencing of trios, mother, father, and the affected child, is advisable [13].

Mutation detection is obviously extremely interesting in oncology where it has potential applications in diagnostics, prognostication, and therapy response prediction [14] (Figure 4). Cancers are caused by specific somatic mutations sometimes occurring in concomitance to predisposing germ line mutations. Somatic mutations can be classified into driver mutations on which cancer growth and development rely and that underlie positive selection during tumor progression and passenger mutations that randomly occur especially in high-grade cancers or at later stages of tumor progression due to genomic instability [15]. In some instances such as melanoma, passenger mutations accumulate due to the carcinogenic effect of the etiological agent itself, UV light in the case of melanoma [16], leading to carcinogen-induced mutations. The knowledge of the driver mutations can be exploited for targeted therapies as well as for the identification of alternative pathways used by the tumor cells for development of resistance to antitumoral drugs. The number of mutations present in a tumor correlates with its malignant potential since highly mutated cancer normally shows a strongly de differentiated phenotype with uncontrolled growth and a higher resistance potential. Yet the full prognostic potential of the mutational pattern has not yet been explored. NGS has shown that structural genomic alterations are much more frequent than previously assumed, and many new translocations leading to fused oncogenes have been identified [17].

As opposed to mutations, variants such as SNPs are not directly related to specific diseases. It is assumed that they are the consequence of occasional mutation followed by fixation in the population in the absence of selective (dis-)advantages. However, they can be associated with diseases where they might modify the disease-causing potential of mutations or environmental factors. It is most likely, although the formal proof is missing, that several SNPs in concert can determine an elevated susceptibility to disease or straightaway cause it [18].

GWAS can identify such multigenic determination of phenotypes among which disease susceptibility. However, the limitation of this is best illustrated by a GWAS on human stature [19]. Twenty SNPs have been identified to be associated with body height, and they elegantly fit into a linear model where the actual stature correlates with the number of SNPs present with their "tall" allele. Yet this model explains with 6 cm only a minor part of the potentially genetically determined variability in human height. Similar results have been obtained in GWAS for breast cancer risk [20]. The need to impute the state of most SNPs due to technical limitations of the widely applied SNP array technology may be overcome through the application of NGS, which can effectively measure the actual state of each single SNP including very rare or private variants.

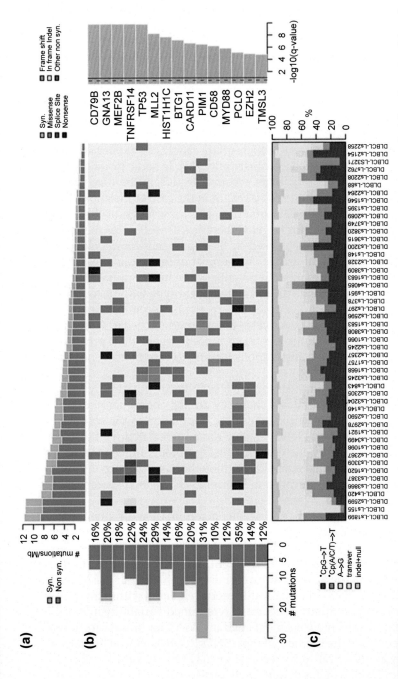

FIGURE 4 Mutation analysis by next-generation sequencing. Significantly mutated genes in 49 patients with diffuse large B-cell lymphoma (DLBCL). (a) The rate of synonymous and nonsynonymous mutations is displayed as mutations per megabase, with individual DLBCL samples ranked by total number of mutations. (b) The heat map represents individual mutations in 49 patient samples, color-coded by the type of mutation. Only one mutation per gene is shown if multiple mutations were found in a sample. (Left) Histogram shows the number of mutations in each gene. Percentages represent the fraction of tumors with at least one mutation in the specified gene. (Right) The 15 genes with the lowest q1-value, ranked by level of significance. (c) Base substitution distribution of individual samples, ranked in the same order as in A. *Figure from Ref. [93] with permission from the National Academy of Sciences, U.S.A.*

Structural Alterations of the Genome

Structural alterations of the genome can be identified by microarrays (aCGH) or NGS. These techniques have revealed that structural variability among healthy subjects is much more extended than previously assumed, and the contribution of such variants to the variable individual susceptibility to specific diseases is only beginning to be fully appreciated [17]. NGS is able to identify translocations, as well as other long-range structural alterations such as inversions, insertions, and deletions [21].

Structural alterations are extremely frequent in cancers. In extreme cases, the whole genome appears to be rearranged, a process called chromothripsis. Other cancers are characterized by a stable genome with hypermutation (amplification) of defined genomic regions, a condition termed kataegis. Chromothripsis is likely a terminal state of genomic instability [22], while kataegis appears to be determined by mutations in genes of the APOBEC family of deaminases [23].

Structural alterations have been associated with a variety of conditions including autism. An inversion polymorphism present mainly in the European population has been associated with increased fertility [24].

MLPA is meanwhile routinely applied in substitution of the much less reliable fluorescence in situ hybridization techniques for the identification of diagnostic or prognostic chromosomal alterations observed, for example, for uveal melanoma, a rare tumor whose metastatic potential is associated with loss of one copy of chromosome 3 and amplification of the long arm of chromosome 8 [25].

EPIGENOMICS

Epigenomic Technologies

Methylation

DNA methylation was originally studied in a genome-wide fashion by restriction landmark genome scanning [26] after bisulfite modification of the DNA. Upon bisulfite treatment, cytosine residues are converted to uracil, but methylated cytosines, 5-methylcytosine, residues remain unaffected. The bisulfite treatment-induced changes in the DNA sequence, therefore, depend on the methylation status of individual cytosine residues and eventually alter restriction sites yielding different patterns for nonmethylated as compared to methylated DNA.

In order to detect methylation events at the single nucleotide level, microarrays containing probes for both the original and the bisulfite-modified sequences complementary to known methylation sites (CpG-islands) have been designed [27]. As an alternative, genome spanning tiling arrays containing probes for the whole genome can be used.

More recently, NGS has been introduced for the analysis of DNA methylation. Sequencing after bisulfite modification of DNA yields genome-wide information on DNA methylation [28].

Histone Modifications

Histone modifications (acetylation, methylation, phosphorylation, ubiquitination of amino acid residues) are the second most studied epigenetic alterations of the genome that contributes to the regulation of gene expression. Histone modifications are studied using specific antibodies for immunoprecipitation of chromatin fragments (nucleosomes). The DNA sequences contained in the precipitated fragments can be analyzed by tiling microarrays (ChIP on chips) [29] or by NGS (ChIP-seq) [9] in a genome-wide manner.

Applications of Epigenomics in Translational Medicine

Epigenomic signatures have been developed and shown to correlate with specific (patho)physiological states of the cell. As an example, a body mass index-associated methylation profile can be cited [30]. Importantly, the methylation patterns identified appear to be stable over time.

TRANSCRIPTOMICS

Transcriptomics, the generation of global gene expression profiles, is the most widely performed genomic technology in basic and translational research. The transcriptome constitutes an intermediate phenotype and is the first step in the information flow from DNA to cellular function. As the "oldest" omic it has yielded impressive results in most fields of biology.

Transcriptomic Technologies

Transcriptome (gene expression) analyses can be performed with qPCR, microarrays, and NGS. The choice of the technology depends on the number of genes to be analyzed. qPCR is best suited for the analysis of limited numbers of genes (<100); microarrays can be customized to analyze subsets of the genome; yet most often, whole genome microarrays are used.

For whole genome analyses, NGS (RNA-seq) [31] is substituting microarrays since RNA-seq yields less biased information on the actual transcriptome, because detection of transcripts is not dependent on predefined probes. The quantitative representation of transcripts is also considered to be more reliable inasmuch as RNA molecules are "counted" through sequencing. However, preparation of the sample for RNA-seq comprises steps that may induce bias and a systematic comparison between these techniques is missing.

ChIP techniques (see above) can also be considered transcriptome-related techniques since they identify gene regulation networks [9,29].

Applications of Transcriptomics in Translational Medicine

Molecular Classification

The most impressive success of genomics has been seen in the field of molecular classification of biological samples, especially in oncology [32]. Molecular classification relies on the analysis of the transcriptome with microarrays or by RNA-seq. The expression values obtained for each gene, the expression profile, constitute a relatively stable characteristic of the each cell or tissue type. It is the molecular correlate of the morphology routinely analyzed by the pathologist but in addition, it delivers information on functional states of the cells. The general profile is mainly determined by the lineage the cells belong to [33], whereas drugs, hormones, growth factors, cytokines, etc., determine more or less extended alterations of subsets of genes. The expression profile appears to be influenced by mutations and polymorphisms much less than expected. In fact, molecular classes (subtypes) identified not necessarily present distinct mutational spectra.

Molecular classification is performed by grouping (clustering) similar samples together. This is obtained through the application of clustering techniques such as hierarchical clustering (HCl) [34]. HCl produces intuitive dendrograms where similar samples are grouped together by branches of a tree whose lengths indicate the degree of similarity. The two main parameters that are chosen for HCl are distance and linkage. Distance is measured by considering each gene as a dimension. The procedure is easy to understand considering two or three dimensions (genes) that can be graphically represented. The extension to more dimensions (actually tens of thousands) is mathematically possible yet difficult to imagine. In the three-dimensional space the concept of distance is straightforward since the location of each sample is defined in the three-dimensional space. There are many alternatives to this "Euclidean" definition of distance, for example, correlation analyses using Pearson or Spearman algorithms. Linkage defines how new samples are joined to existing clusters considering the most distant (complete linkage), the closest (single linkage), or the average (average linkage).

HCl results normally are presented as "heat maps" (Figure 5) where samples and genes are organized in columns and rows so that each square in the diagram represents the expression values of a single gene in a single sample. The actual numeric value is substituted by a color scale where values above the mean are indicated in "hot" colors (most often red), values below the mean in "cold" colors (blue or green), values at the mean in black or white, and missing values in gray. The dendrograms of the

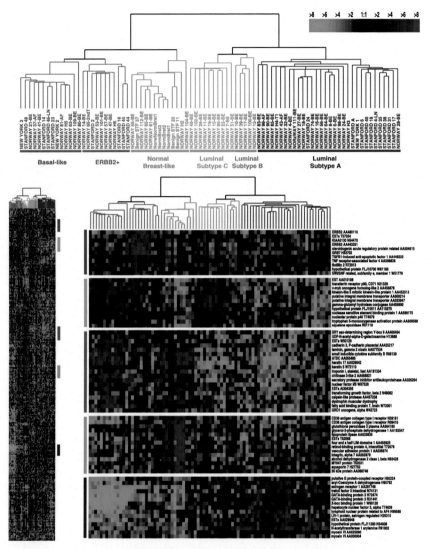

FIGURE 5 **Gene expression profiles for class discovery: identification of breast cancer subtypes.** Human breast cancer samples were analyzed by microarray gene expression profiling. The results are presented as a heat map representing normalized expression values as a color code (red, expression above the mean; green, expression below the mean; black, expression at the mean; gray, data missing). Expression profiles were clustered by hierarchical clustering [2]. The distance of samples is represented by the length of the branches of the tree. (a) Expanded hierarchical tree indicating clusters of breast cancer subtypes, (b) Clustering of the entire expression profile, (c–g) gene expression patterns that identify breast cancer subtypes. Gene expression profiling identifies two main clusters corresponding to estrogen receptor α-positive (on the right) and estrogen receptor α-negative (left) breast cancer cases. The two cancer types are further divided into two or three subtypes, luminal A and luminal B and C for ER+ cases and normal-like, ERBB2+ and basal-like for ER-. *Figure from Ref. [35] with permission. Copyright (2001) National Academy of Sciences, U.S.A.*

samples are added on top, on the left the dendrograms for genes, since HCl can be performed to group together both, samples and genes. This allows for an easy interpretation of the heat map where similar samples and the genes by which they are characterized are recognized by similar distributions of colors.

Molecular classes are clusters with a considerable distance (length of the dendrograms branches) and can be considered as biologically significant if similar clusters are formed when analyzing independent data sets of similar tissues under similar conditions. The relation to clinical and pathological parameters is usually indicated by color bars above the dendrograms where the status of each sample is reported by a color code.

An important example of molecular classification is the identification of molecular subtypes in breast cancer. The known distinction of estrogen receptor α-expressing cancers (ER+), cancers with amplification of the genomic region containing the gene ERBB2 (HER2+), and estrogen receptor α-negative cancers (ER−) was also evident at the level of gene expression that allowed to assign these cases to different clusters. In addition, gene expression profiling allowed for the identification of subtypes with a distinct clinical outcome. This molecular classification has been adopted in the clinics for the distinction between luminal A and luminal B ER+ cancers, with the latter facing an increased risk of relapse [35](Figure 5).

Prognostication

Molecular classes, as observed for breast cancer, can by themselves show a prognostic potential. Complex gene expression data are ideally suited for the definition of prognostic classifiers. Many different procedures to calculate "signatures," sets of genes that are correlated to a specific end point, such as disease-free survival (DFS), relapse-free survival, or overall survival (OS), have been developed and applied to many different cancer types. Prognostic signatures must be developed on large data sets that can be randomly split into a training set on which genes whose expression values correlate with DFS or OS are selected and in a validation set used for testing the actual discrimination power of the signature. In this procedure, a score is calculated for each gene that indicates the extent to which the single gene contributes to the prognostic classifier. The score can be positive if the gene contributes to prolonged DFS or OS and negative if the expression of the gene is associated with an adverse outcome. The score is multiplied with the expression value observed and the addition of the scores of all classifier genes yields a combined score. Classification of new cases is based on this combined score. In the simplest application of this procedure, the median value of the combined score is used as a discriminator and new cases are classified according to the side of the median they fall. During validation, the scores calculated on the training set must be applied without further correction; the validation may not be used to refine the classifier. Before application, the prognostic signature must be tested through external validation on unrelated data sets. Finally, prospective validation best in the context of clinical trials is needed [36].

Again, breast cancers set the stage. Several prognostic classifiers have been developed and are presently in use. The 70-gene signature developed in Amsterdam [37] was the first prognostic signature to be applied followed by the more frequently used Oncotype DX recurrence score [38] that has been developed on the base of microarray gene expression data but has been formulated as a qPCR assay suitable for the analysis of formaldehyde-fixed paraffin-embedded material.

Response Prediction

Predicting the response of a patient to a treatment holds the promise to treat each patient with the drug from which he/she can benefit and avoiding unnecessary side effects and toxicities. Drug toxicities are becoming the more important, the more efficacious the treatments are, since more and more cancer patients survive the neoplastic disease and face long-term toxicity such as cardiotoxicity of the drugs that helped them to survive [39].

Response prediction can be addressed in a similar manner as prognostication; however, microarray gene expression profiles have shown to be less efficient for this purpose. Prognosis is formulated on the base of the molecular profile of the primary tumor that, to a certain extent, already contains the information on the metastatic potential of a tumor. Therapy failure and relapse under treatment are expected to rely to a large extent on acquired molecular features not present in the primary tumor. Therefore, NGS appears to be much more suited for the purpose of response prediction inasmuch as resistance depends on specific mutations [40]. Microarray analyses can identify the activation of escape pathways. The selective pressure operated by anticancer drugs stimulates the growth of minor clones already present in a heterogeneous population of cancer cells or of clones that acquire resistance through novel molecular alterations. The former cannot be identified by expression profiling of the primary tumor where the bulk of the tumor dominates the profile. Deep sequencing can identify such minor populations carrying resistance mutations [41]. Acquired molecular characteristics, mutations, or gene expression alterations can only be identified by analyzing new biopsies.

PROTEOMICS

Proteins regulate response mechanisms and control most of the metabolic activities required for life. In the past, molecular biologists usually analyzed biological mechanisms by studying the role of a single protein, or the interaction between two molecules, rather than all the proteins in the cell. While this reductionist approach has been very successful, it cannot provide the information needed to understand how cellular systems work as a whole, where many competing interactions are taking place between thousands of different components.

The impressive amount of data generated by the genomics revolution is being organized and made accessible in a variety of databases and libraries. These databases define the identity of some of the genes, RNAs, or proteins expressed by a tissue or cell, as well as their structure, function, and macromolecular interactions but all the measurements are in a static manner; on the contrary, many biological processes are dynamic responses to extraneous perturbations (like drugs, disease, and environment). The ability to detect accurately and to quantify all the changes included by a specific perturbation is therefore an essential part of the study of dynamic biological processes.

Proteomics analyzes the whole range of proteins in an organism (proteome), at a particular time, primarily by mass spectrometry (MS) [42,43]. Moreover this approach allows estimating the posttranscriptional modifications (PTMs) and the protein–protein interactions (PPI), data that otherwise are impossible to obtain in any investigation based on DNA or RNA. Indeed, analysis of the proteins rather than the mRNA levels may reflect the functional phenotype of the cells more directly. Moreover, some proportion of the changes at the genome as well as at the transcriptome levels are eliminated by higher regulatory mechanisms. This explains why the relatively young research field of proteome analysis has rapidly become a key technology in biological research. Proteomics, now, rapidly evolves from a discovery-oriented technique to a robust and sensitive quantitative tool to study changes in protein expression and protein modifications in a high-throughput manner [44–48].

Traditionally, the proteins of interest are characterized by targeted approaches such as antibodies, immunoblotting, microscopy, or fluorescence-activated cell sorting. Each of these methods shows only some details of the highly interconnected protein networks. Proteomics based on the latest generation of MS has reached very attractive levels of accuracy, precision, and speed. Recent advances in chromatography and MS have made possible rapid and deep proteomic profiling. New protocols that combine improved sample preparation, chromatographic conditions, and column heater can increase the number of proteins identified across a single liquid chromatography–tandem MS (LC–MS/MS) separation, thereby reducing the need for extensive sample fractionation. This strategy allowed, for example, the identification of the nearly complete yeast proteome (up to 4002 proteins) over 70 min of LC–MS/MS analysis [49]. This makes the characterization of thousands of proteins in a few minutes or the entire proteome of human cells feasible by applying a very limited number of chromatographic fractionations.

Another aspect of great significance is the total independence of the method from specific reagents, offering a hypothesis-free and system-wide analysis. It is possible to define the content, relative abundance, modification states, and interaction partners of proteins in a dynamic and temporal manner on a near-global basis in organelles, whole cells, and clinical samples, providing information of unprecedented detail. These technologies can be applied in a wide array of studies including defining the subcellular locations of proteins in health and disease, connecting cancer genotype to molecular phenotype, unraveling the basis of the innate immune response, identifying the mechanism of action of drug like molecules, and discovering and verifying protein biomarkers of disease.

Operating Principle

Typically, the first step in any proteomics experiment is sample lysis and protein extraction from cells, tissues, or body fluids followed by proteolysis into peptides. Proteins are cleaved into peptides by sequence-specific proteases such as trypsin; the use of other proteases may improve the sequence coverage or help to generate peptides not produced in a standard digestion like the hydrophobic peptides in transmembrane proteins that are usually underestimated. After digestion, the resulting peptide mixtures are cleaned, desalted, and concentrated, commonly in pipette-based devices. Thus, the sample is injected in an ultrahigh-pressure liquid chromatography (UHPLC) combined with a high-resolution MS with very high sequencing speed. During chromatography, peptides are separated in a gradient of aqueous to organic solvent based on their interaction with the hydrophobic (C18) stationary phase. Eluting peptides are ionized by electrospraying at the tip of the column and transferred into the vacuum of the mass spectrometer for direct analysis. The mass spectrometers collect three pieces of information from each peptide: its mass, its ion intensity (MS1), and a list of its fragments (MS2). The peptide mass and fragment masses are used to identify the peptide, whereas the intensity is used for quantification [50] (Figure 6).

Using this approach, several studies have shown that the technology is able to identify ~11,000 expressed protein-coding genes in cultured human cells before saturation of the methodology, indicating that full proteomes can now be sampled, allowing for the exploration of tissue-specific, cell-specific, and developmental-stage-specific protein expression. In addition, the RNA sequences derived from NGS platforms might contain exciting discoveries, such as the presence of new splice variants and even new genes. Based on these opportunities, proteogenomics is developing [51]. This is an area of research at the interface of proteomics and genomics. In this approach, customized protein sequence databases generated using genomic and transcriptomic information are used to help identify

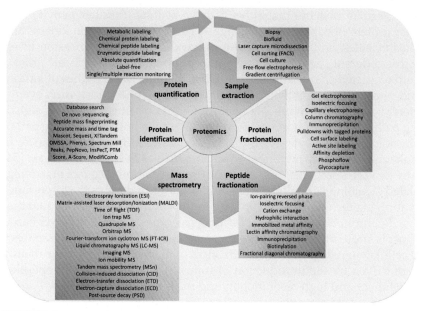

FIGURE 6   Proteomics. This figure depicts the proteomic workflow from sample extraction to protein quantification. For each step in the workflow, the text boxes give examples of commonly used techniques, many of which may be combined in any one study. All featured techniques are discussed in detail in Mallick and Kuster [94] and its supplementary files. *Adapted from Ref. [94] by permission from Macmillan Publishers Ltd.*

novel peptides (not present in reference protein sequence databases) from MS-based proteomic data. The proteomic data can be used to provide protein-level evidence of gene expression and to help refine gene models.

Moreover, like transcriptomics, which allows comparison of mRNA levels between samples, implementation of quantification strategies in the proteomic workflow allows for relative protein quantification with high accuracy. These findings offer new insights into the mechanisms of gene expression control and the relationships between transcription and translation.

Deciphering Posttranslational Regulation (Phosphorylation and Glycosylation)

MS is currently the most powerful tool for studying PTMs in system-wide approaches. PTMs are key regulators of protein activity and involve the reversible covalent modification of proteins by small chemical groups or by the cleavage of proteases [52].

Protein phosphorylation, probably the best-studied PTMs, is implicated in a variety of cellular processes, spanning from proliferation and differentiation to apoptosis. Site-specific phosphorylation events can function as molecular switches that either activate or inhibit protein activity. Cellular protein phosphorylation is tightly controlled by protein kinases and phosphatases, and as these enzymes have differential expression levels across different cell types, protein phosphorylation is a dynamic process with restricted spatial and temporal distribution. The activities of kinases and phosphatases are themselves fine-tuned by phosphorylation events, thereby creating a complex regulatory pattern by interconnecting signaling pathways. Phosphorylation events have been implicated in the pathophysiology of several severe diseases, such as cancer. For instance, in leukemia, activating mutations in kinases such as flt3 [53] and bcr–abl [54] are often the oncogenic drivers of cell transformation. The fact that deregulated signaling is a hallmark of many diseases highlights the importance of applying techniques that allow for rapid, comprehensive, and quantitative determination of disease phosphoproteomes. Quantitative MS-based phosphoproteomics is currently the most powerful technique for analysis of cellular signaling networks. Advances of the methodology have mainly been driven by the introduction of robust methods for phosphopeptide enrichment and high-resolution hybrid mass spectrometers.

Another field of growing importance is the characterization of carbohydrates and in particular the glycoproteins. Posttranslational modification of proteins with glycans to form glycoproteins is a common biological process. Glycans in glycoproteins are involved in a wide range of biological and physiological processes including recognition and regulatory functions, cellular communication, gene expression, cellular immunity, and growth and development. Aberrant glycosylation of proteins is connected to cancer progression and invasion and metastasis. Glycan functions are often dependent on the structure of the glycans attached to the protein. Glycans are covalently attached to proteins primarily through two structural motifs. They can be attached to the amide group of an asparagine, referred to as "N–linked glycans" or attached to proteins through the hydroxyl group on serine or threonine, referred to as "O–linked glycans." The biological activity and function of N–linked glycans are better studied than O–linked glycans; however, both types of glycans are investigated as biomarkers, in order to understand changes related to complex organelle development, and as part of therapeutic protein drug development with strong indication that efficacy is affected by glycosylation. This particular PTM is difficult to handle, due to the great structural variability of glycans and a large amount of fragments that must be interpreted and relocated in a specific pattern. Research, however, has taken an incredible leap forward, thanks to the new generation of mass spectrometers that allow multiple and different fragmentation (MS2) in parallel, without losing the scanning speed at the expense of data collection. Moreover, the identification of diagnostic peaks, uniquely associated to glycoPTMs, has enabled the development of highly selective methods for the analysis of glycoproteins.

Clinical Applications

As stated above, the immense complexity of the molecular and cellular processes hinders the identification and interpretation of the molecular causes of diseases. However, a very promising field is the combination of different technologies in order to understand drug resistance in cancer therapy. Tumors are driven by gene mutations that result in constitutively active signaling pathways controlling proliferation and growth. Inhibiting only individual nodes can select alternative routes that lead to drug resistance. Instead, a coordinated application of two factors, such as two different drugs or a drug and interventions on life style and diet, can deliver a synergistic effect, preventing the rewiring of proliferative pathways [55].

Clinical Biomarkers

The evaluation of biomarkers has become one of the focal points in clinical studies prompted by a need for better markers for most diseases due to the limited performance of those currently used, especially in the field of oncology [56]. The process of evaluating a large number of biomarker candidates, typically derived from discovery studies, relies on a high-throughput approach, which will rapidly materialize a small panel of markers that can be translated into a routine clinical assay. An MS-based approach provides the exquisite selectivity and sensitivity to perform such a task. It allows the peptides (obtained by enzymatic digestion of the proteins) to be rapidly detected and measured in body fluids to determine the protein concentrations and their differential expression in various pathophysiological conditions.

At present, one of the rate-limiting steps in the development of new biomarkers is the high-throughput verification of the numerous candidates obtained from proteomic discovery experiments, genomic screening, or literature mining, which typically require a number of distinct steps. In contrast, an MS-based strategy is fast with respect to method development, associated with low-cost reagents (isotopically labeled synthetic peptides), and is easily scalable.

The Clinical Proteomic Tumor Analysis Consortium (CPTAC)

The information generated in a typical proteomics experiment can be organized in three different levels: (1) raw data; (2) processed results, including peptide/protein identification and quantification values; and (3) the resulting biological conclusions. Technical and/or biological metadata can be independently provided for each level. These three categories enable the classification of the existing MS proteomics repositories according to their level of specialization. Compared to other data-intensive fields such as genomics, deposition, and storage of original proteomics, data in public resources is less common. This is regrettable since proteome studies are usually more complex than genome studies. In fact, data interpretation in proteomics can be considerably more complex than in genomics due to the wide variety of analytical approaches, bioinformatics tools and pipelines, and the related statistical analysis.

The clinical proteomic tumor analysis consortium (CPTAC, http://proteomics.cancer.gov/programs/cptacnetwork) of the National Cancer Institute is a comprehensive and coordinated effort to accelerate the understanding of the molecular basis of cancer through the application of robust, quantitative proteomic technologies, workflows, and databases [57]. This initiative can be seen as the integration of the different approaches described above. CPTAC is a network of proteome characterization centers that coordinates and conducts research and data sharing activities in order to comprehensively examine genomically characterized cancer biospecimens. Importantly, CPTAC data with accompanying assays and protocols are made publicly available. This multidisciplinary (proteomics, genomics, bioinformatics, experimental design, statistics, cancer biology, and oncology) consortium identifies proteins that result from changes in cancer genomes and their related biological processes. Understanding these functional changes at the protein level is the next step in better defining the molecular mechanisms of cancer, such as the deregulated signaling pathways responsible for tumorigenesis and metastasis. Therefore, CPTAC will analyze changes in protein expression, their posttranslational modifications and variations, as well as those in protein–protein interaction and signaling networks responsible for the pathological state of cancer.

METABOLOMICS

Overview of Metabolomics, History, and Definition

The development of modern metabolomics has become possible in the 1960s through improved chromatographic separation techniques that allowed for the analysis of a large number of compounds obtained from complex samples [45]. The term "metabolome" was coined in the 1990s by Oliver et al. referring to "the complete set of metabolites/low-molecular weight intermediates, which are context dependent, varying according to the physiology, developmental or pathological state of the cell, tissue, organ or organism" [58]. These molecules are organic compounds such as sugars, lipids, amino acids, nucleotides, or others. The analysis of qualitative and quantitative variations of all metabolites in a biological system is called metabolomics. This method, while not yet defined as metabolomics, was devised by A. Robinson and L. Pauling in 1970, through their studies on body fluids analyzed by chromatographic techniques [59]. The term metabolomics refers to metabolic profiles of living organisms [58], whereas the term metabonomics is used as an alternative to metabolomics [60].

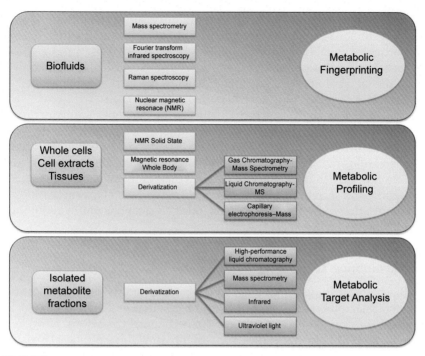

FIGURE 7 **Metabolomics.** The diagram shows three metabolomics approaches, metabolic fingerprinting, metabolic profiling, and metabolic target analysis with the associated technologies.

The metabolome is the result of gene and protein expression in response to external stimuli, such as environmental changes, nutritional stimuli, drugs, and genetic alterations [61]. While the genome is representative of the potential of the system under investigation and the proteome is the part of this potential that is actually expressed, the metabolome represents instant by instant the current status of the cell, tissue, organ, or organism. Therefore, the metabolome provides a snapshot of the biological system providing a large amount of information to determine the role of metabolic pathways in the feedback regulation of gene expression and protein [58,60] (Figure 7).

Metabolomics is based on many different analytical approaches [62] among which (Figure 7) are:

- metabolic fingerprinting, which measures a subset of the whole profile with little differentiation or quantification of metabolites [63].
- metabolic footprinting, which analyzes the metabolic exchange between the biological system under investigation and its environment (mainly indicated for in vitro cell system analysis) [64].

- metabolic profiling, which is the quantitative study of a group of metabolites, known or unknown, within or associated with a particular metabolic pathway [65].
- target isotope-based analyses, which focus on a particular segment of the metabolome by analyzing only a few selected metabolites that comprise a specific biochemical pathway [66].
- metabolic flux analysis, which quantifies the metabolic pathway fluxes in vivo and in vitro, using labeled substrates, in specific locations. This approach allows to follow the effect and the connections between the various metabolic pathways in cells, tissues, organs, animals, and humans. It is, thus, possible to obtain a map of metabolic fluxes (as uptake and secretion) of a biological system. These maps show the distribution of the flows of the anabolic and catabolic pathways in the metabolic network [67].

Metabolomics Technology

Biochemical studies on metabolism have allowed for the identification of molecular processes involved in the regulation of different biological systems in relation to a wide range of experimental conditions. However, conventional metabolic studies focused on single or specific classes of metabolites, macromolecules, or metabolic processes that were considered separately. This type of approach presents obvious limits in the extrapolation of the results to an understanding of the real functioning of a biological system in its entirety. Complexly organized biological systems dispose of alternative ways to determine the flow of materials and the energy necessary for their operation. The identification of the metabolic pathways that lead from a starting substrate to the final products and the description of the enzymes involved are not straightforward. For this reason, it is not possible to define a technique of choice for metabolomics studies [62].

The most important analytical metabolomic techniques are spectrometric techniques (infrared (IR), near-infrared (NIR), ultraviolet (UV) spectrometry), nuclear magnetic resonance (NMR), and mass spectrometry (MS). These techniques usually require complex sample manipulations and long analysis time and have only low sensitivity so that not all metabolites can be quantified. The resolution, sensitivity, and selectivity of these technologies can be upgraded by matching them with separation systems such as gas chromatography (GC), liquid chromatography (LC), and capillary electrophoresis (CE) (Figure 7).

NMR Spectroscopy-Based Metabolomics

The NMR spectroscopy is a powerful tool for metabolic studies and has gained great importance in the biomedical field with regard to the functional characterization of various diseases. It can be applied both in vitro and in vivo and is able to simultaneously quantify many metabolites in the micromolar concentration range within a sample [62]. This analytical technique uses the chemical characteristics of the atoms, such as the chemical shift, coupling, and spin relaxation times, to identify the different metabolites. Nuclear isotopes (nuclides) used in NMR are ^{1}H, ^{13}C, and ^{31}P. Naturally occurring isotopes or enriched isotopes can be used. NMR is a nondestructive and noninvasive technique, but less sensitive than MS. The noninvasiveness makes it suitable for time course analyses on the same biological sample. The metabolism of substrates marked with nuclides can be followed within the biological sample in order to identify metabolic pathways.

Mass Spectroscopy, GC–MS, LC–MS-Based Metabolomics

Most frequently, MS is associated with various chromatographic techniques for metabolomics. Resolution relies on the chromatographic separation of metabolites that are then detected with MS. This is an economic, reproducible, and sensitive technique, but some metabolites, such as volatile or thermolabile compounds, cannot be detected with this method. Samples must therefore be derivatized before separation. Derivatization becomes, however, the limiting step of this analytical technique, since it destroys the sample and has only low productivity. These limits are overcome by LC and other analyzers, such as the time-of-flight (TOF). TOF consists in the ionization of molecules and their spreading in a cathode tube where their TOF, which is proportional to their mass, is measured.

Metabolomic Data Analysis

Just like the other omics, metabolomics generates complex data sets that can be analyzed through the comparison of specific classes such as healthy and diseased tissues or treated and untreated samples. In alternative, complex data can be analyzed in an unsupervised manner for the identification of patterns or classes within complex sample collections. Essentially, the analytical workflow follows the same scheme as for the other omics. Special attention must be devoted to the fact that the metabolome is the less stable and the most easily perturbed intermediate phenotype of the cell.

Applications of Metabolomics in Translational Medicine

The constant development and updating of analytical techniques and diagnostics has enabled the wide use of metabolomics for diagnostic and prognostic purposes. It allows a systems approach on the biological materials available for diagnostics, such as blood, urine, feces, saliva, tissue, and even breath [68]. Yet, metabolomics has an important application in vivo through spectroscopic techniques: PET imaging that allows for the localization of primary tumors and metastases through the analysis of accumulation of radioactive tracers for metabolic substrates such as [18]fluorodeoxyglucose [69].

Metabolomics is widely used for studies of drug discovery [70] and biomarker identification. Some metabolites are, in fact, already clinically relevant [71]. The metabolomic determination of metabolite concentrations and metabolic fluxes is able to distinguish precancerous from cancerous states and to predict and monitor drug responses [72].

As for the other omics, metabolomics data can be mined for the recognition of patterns for the identification of metabolic profiles or signatures. This has been described for many types of tumors, such as head and neck, lung, breast, ovarian, leukemia, prostate, and thyroid [73]. The most important example is on breast cancer; in this study, there were 30 metabolites identified whose concentration changes allow to discriminate between healthy tissue and tumor [74].

Several metabolomic studies have addressed chronic diseases such as Alzheimer [75], hypertension [76], cardiovascular disease [77,78], diabetes [79,80], and obesity [81].

There are also numerous studies on the association between nutrition and metabolism, microbiome, and microbiota. Metabolomics yields insight into the complex mechanisms that regulate the host–microbe interaction [82,83]. Nutritional metabolomics is used to predict the effect of diet, for example, in malnutrition in children [84] or in necrotizing enterocolitis [85].

BIOINFORMATICS

Bioinformatics is an essential element of the omics sciences, since the complex data generated cannot be analyzed without the help of informatics tools. Bioinformatics started with sequence alignment analyses and the generation of nucleotide sequence databases and has evolved through the generation of complex tools for data acquisition, preprocessing (background correction, normalization, etc.), statistical evaluation, functional analyses, data integration, and data representation.

Hardware

Omics generates an ever-growing demand of computational resources. While the first-generation omics applications mainly required potent processing units, the introduction of NGS has generated a demand of storage capacity since whole-genome sequencing and RNA-seq data generate gigabytes of data for each single sample analyzed. More recently, this is being addressed by cloud computing but still, a bioinformatics department must be equipped with powerful storage and computation resources. Hardware requirements will continue to grow in the light of the expected increase in human genomes to be sequenced.

Software

There is a very vivid development of software for the omics sciences with software dedicated to the machines that generate the data, commercial software suites for data analysis, laboratory information management systems, and open source solutions generated by the scientific community that are freely available on the Web. Here, we briefly review only the latter. The most important software is Bioconductor [86] (http://www.bioconductor.org/) a library of "packages" each one designed by a scientist for the solution of a specific bioinformatic problem. Bioconductor is running on the background of R, a program for object-oriented statistical computation (http://www.r-project.org/) that runs on all platforms. Packages contain the algorithms needed to run the analyses together with documentation on the functions of the program. Bioconductor is at present the most comprehensive and the best updated bioinformatic software. For many applications, various solutions are present and the user is enabled to select those best suited for the specific problem. Its use is, however, limited to dedicated personnel since the command-based operation is not suited for occasional use by nonexperts. Bioconductor also contains packages dedicated to proteomics but comprehensive proteomics analysis tools can be found in the freely available MaxQuant [87] and Perseus software [88] (http://www.maxquant.org/).

Data Repositories

Several data repositories have been developed for the storage of omics data. The most important ones are Gene Expression Omnibus (http://www.ncbi.nlm.nih.gov/geo/) run by the National Center for Biological Information, USA, and the ArrayExpress (http://www.ebi.ac.uk/arrayexpress/) run by the European Bioinformatics Institute, UK. Many additional, often more focused, repositories are also available.

Systems Biology

Systems biology is the computational and mathematical modeling of complex biological systems (definition by Wikipedia). Systems biology approaches try to analyze complex biological data without a priori data reduction. The results of these analyses are most often graphically represented by bi- or tridimensional networks where the nodes represent genes or proteins, and the edges identify some type of relation between them, such as physical interactions or coregulation. Software useful for this purpose is, for example, Cytoscape (http://www.cytoscape.org/) applied to network data integration, analysis, and visualization.

The ultimate goal of systems biology is the interpretation of biological data at a systems level. However, at present the main application of the tools developed is the prioritization of biological molecules or pathways for downstream analyses and validation.

Most approaches to data interpretation rely on gene and protein annotations. Gene Ontology (http://geneontology.org/) is a project aimed at gathering biological information in a gene-centered ontology where each gene is associated with a hierarchically organized dictionary of biological processes, molecular functions, and cellular components.

Based on gene ontology, gene or protein lists obtained through omics can be interrogated for the enrichment of specific terms. This is simply performed by the analysis of the frequency of genes annotated with a specific term in the list of genes identified in comparison to the totality of genes in the databank. Perhaps the most comprehensive tool for this type of analysis is Enrichr (http://amp.pharm.mssm.edu/Enrichr/) [89] that integrates almost all types of high-throughput analyses from microarray screenings to ChIP-seq that yield gene-anchored data. Bioprofiling (http://bioprofiling.de/) allows for similar analyses and the integration with Cytoscape for graphical display of the results. HEFalMp (http://hefalmp.princeton.edu/) [90] delivers virtual interaction data based on gene expression and physical interaction that are evaluated using a naïve Bayesian classifier.

Mutation screenings require downstream analyses for developing hypotheses on functional consequences of specific mutations. A wide variety of tools is available. PolyPhen (http://genetics.bwh.harvard.edu/pph2/) yields estimations of the effect of mutations based on analyses of mutations with known functional consequences. canSAR (https://cansar.icr.ac.uk/) helps to verify whether a mutation indicates a specific drug to which the patient might respond through a multifaceted analysis of protein and drug databanks. DrugDB (http://www.drugbank.ca/) yields similar information.

Bioinformatics is a highly dynamic field of investigation and it is impossible to cite all useful approaches. The scientific journal *Nucleic Acids Research* publishes each year a database issue [91] and a compilation of more than 1000 bioinformatic tools is available at the Bioinformatics Links Directory (http://bioinformatics.ca/links_directory/).

Bioinformatics Workflow

High-Throughput Techniques

Invariably, genomics, proteomics, and metabolomics lead to the generation of large data sets. The main analytical approaches to mine the information contained in these data are similar regardless of the specific technique applied.

The workflow starts with data acquisition followed by background correction needed to subtract noise from the data, a fact achieved under the assumption that noise follows a normal distribution, whereas the specific signals show an exponential distribution. Many approaches need a normalization step in which any variability that has been introduced during the experimental procedure can be corrected.

Statistical analyses for the identification of significant alterations of the "-ome" analyzed can be divided into two classes: supervised analyses that start with preexisting knowledge on the samples to be compared and unsupervised analyses aimed at identifying so far unknown classes.

The most frequently used approaches for the latter are clustering methods that group together ("cluster") samples with a similar molecular profile. Similarity is calculated by a number of different methods that also yield a measure of similarity or "distance" on the base of which the samples are clustered using various agglomeration ("linkage") methods. The output, most often represented as a heat map where the original numeric values are represented by colors after normalization (for example, green for "downregulated," red for "upregulated," and black for invariant genes or similar). These heat maps often generate easily interpretable figures for the intuitive identification of molecular classes that are more robust and more distant from each other. An alternative or complementation to clustering methods is principle component analysis, a method that constructs vectors in the multidimensional space that represent as much of the variability as possible to be represented in two- or three-dimensional plots where molecular classes can be identified as groups of samples that occupy distinct positions in the space.

Supervised analyses can be divided into class comparison and class prediction analyses; the former compares two or more classes in order to identify significantly differentially expressed molecular items. The statistics to be applied must take into account that a high number of parameters are analyzed; simple t-statistics is therefore not adequate. Bootstrapping analyses such as "Significance Analysis for Microarrays" are widely used for this scope.

Prediction analyses try to classify new samples based on the classes they belong to. These approaches need a training set on which to develop a classifier and a validation set on which to test it, followed by external validation on unrelated samples. There are a large number of classification methods all of which distinguish two steps: feature selection, where the features (genes, proteins, or similar) whose experimental values can actually distinguish the classes are selected, and the construction of the actual classifier. In the simplest cases, a line in the multidimensional space that can separate the classes is used (linear discriminant analysis) or the distance of the samples belonging to the classes from the mean of all the samples is calculated and new samples are classified according to their distance from this mean (centroid classification). The shrunken centroid method combines feature selection and classifier building in an iteration of steps where the minimum of features for correct classification is identified. More complex classification methods rely on neural networks or other machine learning approaches.

Next-Generation Sequence Data

The analysis pipeline for NGS data starts with the quality control of the raw sequences obtained. The sequences must be aligned to the reference genome in order to obtain a call for discordant positions. Sequences obtained by NGS frequently show artifacts. Therefore, only mutations detected in several sequences can be considered reliable. Each genomic segment should therefore be sequenced at least 30 times with higher coverage rates increasing the reliability. Once unreliable (low-frequency) sequences are filtered, the variant call is made. Variants may correspond to know single nucleotide variants (SNV) or not. In the latter case, they might correspond to de novo mutations. In the case of potentially somatic mutations, as, for example, in cancer samples, a comparison with the germ line sequence (most frequently obtained from peripheral blood monocytes) must be made. Further filtering can distinguish mutations in coding or noncoding regions, silent (with no effect on the protein sequence), nonsilent (amino acid changing) mutations, or splice-site-affecting mutations. In addition to SNVs, insertions and deletions (INDELs) and, according to the specific approach, translocations and the corresponding fusion transcripts (in the case of RNA-seq) must be identified.

NGS data analysis is rapidly evolving and extensive software is available from Bioconductor (www.bioconductor.org).

CONCLUSIONS

Omics have become an integral part of translational research and are delivering a wealth of information on many human diseases. In most instances, omics are applied in an intermediate development phase where the molecular players are identified. The development of clinical applications usually still relies on other less complex technologies able to analyze single genes, proteins, or metabolites or small sets of them in a highly reliable manner. In the case of NGS, in many cases the analysis of the single mutation identified can be sufficient. Yet especially in the case of complex diseases, the sequences of large sets of genes can be useful to guide the clinical decision. In the light of the ever-growing knowledge on genotype–phenotype relations and dropping sequencing costs, the introduction of whole-genome sequencing as a standard measure for each patient is foreseeable. Bioinformatics and especially systems biology will become more and more important with the potential to shift our understanding from the level of single genes or proteins to a true system level addressing the biological complexity and its pathological perturbations. Germ line genetics will be exploited to better estimate the individual disease risk and targeted prevention strategies can be initiated early in life with a potential to increase life span and quality of life. A basic understanding of these technologies, their characteristics, their limits, and their pitfalls is therefore important for every physician and every biomedical researcher.

Glossary

Amplicon A defined stretch of chromosomal DNA that undergoes amplification

Amplification Genetic mechanism by which the copy number of a gene is increased in amount above the normal diploid status. Chemically, it is possible to increase the copy number of a nucleic acid stretch by the polymerase chain reaction.

Biallelic variants Variants present in the population with two alleles of the four theoretically possible variants

cDNA Double-stranded DNA synthesized from a messenger RNA template

Chromatin immunoprecipitation (ChIP) Chromatin immunoprecipitation, a molecular analysis method where the DNA sequences recognized by DNA-binding proteins are identified through precipitation of the protein–DNA complex using specific antibodies.

Chromothripsis A catastrophic event leading to thousands of clustered chromosomal rearrangements in confined genomic regions in one or a few chromosomes, known to be involved in cancer and congenital diseases.

CpG island Genomic region of typically 300–3000 bp length with enhanced occurrence of CG dinucleotides, frequently occurring in transcription regulatory regions of genes

cRNA Synthetic RNA produced by transcription from a specific single-stranded DNA template

Domain knowledge Potential knowledge intrinsic to data collections and repositories not necessarily known to any physical person

Exon The part of the gene that will be encoded, that remains present within the final mature RNA, after introns have been removed by RNA splicing

Fluorescence in situ hybridization (FISH) Molecular technique for the identification of the position and the number of specific genes within the genome or the transcripts derived there through hybridization with a DNA or RNA probe that is labeled with a fluorescent chromophore

Gene conversion Loss of one allele and the duplication of the other leading to uniparental disomy or isodisomy

Genome-wide association studies (GWAS) Analysis of the association of polymorphisms (SNPs) with specific traits in cohorts of individuals at the genome level typically involving hundreds of thousands of SNPs

Genotype Genetic composition of an organism or a trait of an organism

Germ line mutation Mutation occurring in the germ line pool of genes

Germ line The pool of genes that is transmitted from one organismic generation to the next

Glycoproteomics Analysis of the whole range of glycoproteins, with special emphasis on membrane or secreted proteins.

Glycosylation Glycosylation (see also chemical glycosylation) is the reaction in which a carbohydrate, i.e., a glycosyl donor, is attached to a hydroxyl or other functional group of another molecule (a glycosyl acceptor).

Haploblock Genomic sequence containing a variable number of polymorphisms that are usually inherited together due to the lack of recombination within the block

Heat map Graphical representation of numerical data where normalized data are colored according to their distance from the mean value

Insertion Addition of one or more nucleotide base pairs as well as the insertion of a larger sequence into a DNA sequence or chromosome

Inversion A rearrangement in which a chromosome segment is clipped out, turned upside down, and reinserted back into the chromosome

Isodisomy Uniparental disomy

Kataegis Hypermutation occurring in restricted chromosomal regions during carcinogenesis of some cancer subtypes

Metabolic fingerprinting Characterizing samples on the basis of their origin and/or biological relevance

Metabolic or metabolite profiling Analysis of a group of metabolites that are part of a particular group or class (e.g., sugars, lipids), or linked to a specific pathway

Metabolite Small (usually less than 1500 molecular weight) biological compounds involved in biochemical processes and pathways

Metabolome A collection of all metabolites observed in a biological system under a given set of conditions

Metabolomics The analysis, quantification, and interpretation of metabolite levels in biological samples with the view to characterizing a metabolome

Metabonomics The study of changes in metabolite levels in response to drugs, diseases, or toxicity, usually used when determining such changes using NMR

Methylation The modification of a strand of DNA after it is replicated, in which a methyl (CH_3) group is added to any cytosine molecule that stands directly before a guanine molecule in the same chain

Microarrays Miniaturized arrays of biological probes deposited on a solid surface such as glass or silicon where specific probes of known sequence and reactivity are located in predetermined positions.

Mosaicism Denotes the presence of two or more different genotypes in one individual. Mosaicism can result from various mechanisms including chromosome nondisjunction and anaphase but can also result from a mutation during development which is propagated to only a subset of the adult cells.

MS/MS or MS2 An MS/MS or MS2 spectrum consists of m/z versus intensity of fragment ions derived from peptide precursors that had been isolated by m/z before fragmentation using collisions with an inert gas

MS1 An MS1 spectrum consists of the mass-to-charge ratios (m/z) and signal intensity (charges/second) of the collected intact peptide ions

PCR A molecular biology technology used to amplify a single copy or a few copies of a piece of DNA across several orders of magnitude, generating thousands to millions of copies of that particular DNA sequence

Phenotype The ensemble of all the measurable or observable traits of an organism

Phosphoproteomics Analysis of cellular signaling networks, involving the study of phosphorylated proteins, kinases, and phosphatases.

Phosphorylation The transferring of phosphoryl group from a donor to the recipient molecule

Polymorphism Genetic variant of variable length (SNP) present in the population with various frequencies

Probe An RNA or DNA molecule that specifically anneals with a complementary nucleic acid to be detected or analyzed

Proteogenomics Protein sequence databases generated using genomic and transcriptomic information used to identify novel peptides from mass spectrometry-based proteomic data

Proteomics Analysis of the whole range of proteins in an organism at a particular time

Signature Group of molecular items (genes, proteins) that distinguish molecular classes

Somatic mutation Mutation that hits the genome of a cell outside the germ line. By definition it cannot be transmitted from one organismic generation to the next

Subtypes Molecular classes that further differentiate histologic tumor types

Translocation Chromosome rearrangement that results in a fusion of two chromosomal segments that are not normally attached with each other

Ubiquitination Ubiquitination is an enzymatic, protein posttranslational modification process in which the carboxylic acid of the terminal glycine from the diglycine motif in the activated ubiquitin forms an amide bond to the epsilon amine of the lysine in the modified protein

Uniparental disomy A situation in which either both chromosomes or part of chromosomes in a pair are from one parent only. It can be the result of heterodisomy, in which a pair of nonidentical chromosomes are inherited from one parent in an early stage of meiosis (uniparental disomy) or isodisomy, in which a single chromosome from one parent is duplicated (a later stage in meiosis).

LIST OF ACRONYMS AND ABBREVIATIONS

aCGH Array-based comparative genome hybridization
C18 Hydrophobic stationary phase
cDNA Complementary DNA
ChIP Chromatin immunoprecipitation
CPTAC The Clinical Proteomic Tumor Analysis Consortium
cRNA Complementary RNA
ER Estrogen receptor
GWAS Genome-wide association studies
HCl Hierarchical clustering
lncRNA Long noncoding
LC Liquid Chromatography
MLPA Multiplex ligation dependent PCR amplification
mRNA Messenger RNA
MS Mass spectrometry
ncRNA Noncoding RNA
NGS Next-generation sequencing
PCR Polymerase chain reaction
piRNA Piwi-interacting RNA
PPI Protein–protein interactions
PTMs Posttranscriptional modifications
qPCR Quantitative polymerase chain reaction, also named real-time PCR
miRNA Micro RNA
SNP Single nucleotide polymorphisms
UHPLC Ultrahigh-pressure liquid chromatography

References

[1] Erlandson A, Samuelsson L, Hagberg B, Kyllerman M, Vujic M, Wahlstrom J. Multiplex ligation-dependent probe amplification (MLPA) detects large deletions in the MECP2 gene of Swedish Rett syndrome patients. Genet Test Winter 2003;7(4):329–32.
[2] Grada A, Weinbrecht K. Next-generation sequencing: methodology and application. J Invest Dermatol 2013;133(8):e11. [Research Techniques Made Simple].
[3] Pinkel D, Albertson DG. Array comparative genomic hybridization and its applications in cancer. Nature Genet 2005;37.
[4] Sale MM, Mychaleckyj JC, Chen WM. Planning and executing a genome-wide association study (GWAS). Methods Mol Biol 2009;590:403–18.
[5] Metzker ML. Sequencing technologies - the next generation. Nat Rev Genet January 2010;11(1):31–46.
[6] Choi M, Scholl UI, Ji W, Liu T, Tikhonova IR, Zumbo P, et al. Genetic diagnosis by whole-exome capture and massively parallel DNA sequencing. Proc Natl Acad Sci November 10, 2009;106(45):19096–101.
[7] Flaherty P, Natsoulis G, Muralidharan O, Winters M, Buenrostro J, Bell J, et al. Ultrasensitive detection of rare mutations using next-generation targeted resequencing. Nucleic Acids Res January 2012;40(1):e2.
[8] Ozsolak F, Milos PM. RNA sequencing: advances, challenges and opportunities. Nat Rev Genet 2011;12(2):87–98.
[9] Park PJ. ChIP-seq: advantages and challenges of a maturing technology. Nat Rev Genet 2009;10(10):669–80. http://dx.doi.org/10.1038/nrg2641.
[10] Tringe SG, von Mering C, Kobayashi A, Salamov AA, Chen K, Chang HW, et al. Comparative metagenomics of microbial communities. Science April 22, 2005;308(5721):554–7.
[11] Koboldt DC, Steinberg KM, Larson DE, Wilson RK, Mardis ER. The next-generation sequencing revolution and its impact on genomics. Cell September 26, 2013;155(1):27–38.
[12] Yang Y, Muzny DM, Reid JG, Bainbridge MN, Willis A, Ward PA, et al. Clinical whole-exome sequencing for the diagnosis of Mendelian disorders. N. Engl J Med 2013;369(16):1502–11.
[13] DePristo MA, Banks E, Poplin RE, Garimella KV, Maguire JR, Hartl C, et al. A framework for variation discovery and genotyping using next-generation DNA sequencing data. Nat Genet 2011;43(5):491–8.
[14] Kilpivaara O, Aaltonen LA. Diagnostic cancer genome sequencing and the contribution of germ line variants. Science March 29, 2013;339(6127):1559–62.
[15] Campbell PJ, Pleasance ED, Stephens PJ, Dicks E, Rance R, Goodhead I, et al. Subclonal phylogenetic structures in cancer revealed by ultradeep sequencing. Proc Natl Acad Sci USA September 2, 2008;105(35):13081–6.
[16] Krauthammer M, Kong Y, Ha BH, Evans P, Bacchiocchi A, McCusker JP, et al. Exome sequencing identifies recurrent somatic RAC1 mutations in melanoma. Nat Genet September 2012;44(9):1006–14.
[17] Liu B, Conroy JM, Morrison CD, Odunsi AO, Qin M, Wei L, et al. Structural variation discovery in the cancer genome using next-generation sequencing: computational solutions and perspectives. Oncotarget March 20, 2015;6(8):5477–89.
[18] van der Harst P, Zhang W, Mateo Leach I, Rendon A, Verweij N, Sehmi J, et al. Seventy-five genetic loci influencing the human red blood cell. Nature December 20, 2012;492(7429):369–75.
[19] Weedon MN, Lango H, Lindgren CM, Wallace C, Evans DM, Mangino M, et al. Genome-wide association analysis identifies 20 loci that influence adult height. Nat Genet May 2008;40(5):575–83.
[20] Michailidou K, Beesley J, Lindstrom S, Canisius S, Dennis J, Lush MJ, et al. Genome-wide association analysis of more than 120,000 individuals identifies 15 new susceptibility loci for breast cancer. Nat Genet April 2015;47(4):373–80.

[21] Chen W, Kalscheuer V, Tzschach A, Menzel C, Ullmann R, Schulz MH, et al. Mapping translocation breakpoints by next-generation sequencing. Genome Res 2008; 18(7):1143–9.

[22] Forment JV, Kaidi A, Jackson SP. Chromothripsis and cancer: causes and consequences of chromosome shattering. Nat Rev Cancer October 2012;12(10):663–70.

[23] Lada AG, Dhar A, Boissy RJ, Hirano M, Rubel AA, Rogozin IB, et al. AID/APOBEC cytosine deaminase induces genome-wide kataegis. Biol Direct 2012;7:47. discussion.

[24] Stefansson H, Helgason A, Thorleifsson G, Steinthorsdottir V, Masson G, Barnard J, et al. A common inversion under selection in Europeans. Nat Genet February 2005; 37(2):129–37.

[25] Damato B, Dopierala JA, Coupland SE. Genotypic profiling of 452 choroidal melanomas with multiplex ligation-dependent probe amplification. Clin Cancer Res Off J Am Assoc Cancer Res December 15, 2010;16(24):6083–92.

[26] Ando Y, Hayashizaki Y. Restriction landmark genomic scanning. Nat Protoc 2007;1(6): 2774–83. http://dx.doi.org/10.1038/nprot.2006.350.

[27] Weber M, Davies JJ, Wittig D, Oakeley EJ, Haase M, Lam WL, et al. Chromosome-wide and promoter-specific analyses identify sites of differential DNA methylation in normal and transformed human cells. Nat Genet 2005;37(8):853–62. http://dx.doi.org/10.1038/ng1598.

[28] Hurd PJ, Nelson CJ. Advantages of next-generation sequencing versus the microarray in epigenetic research. Brief Funct genomics 2009. June 17, 2009.

[29] Weinmann AS, Yan PS, Oberley MJ, Huang TH, Farnham PJ. Isolating human transcription factor targets by coupling chromatin immunoprecipitation and CpG island microarray analysis. Genes Dev January 15, 2002;16(2):235–44.

[30] Feinberg AP, Irizarry RA, Fradin D, Aryee MJ, Murakami P, Aspelund T, et al. Personalized epigenomic signatures that are stable over time and Covary with body mass Index. Sci Transl Med 2010;2(49):49ra67.

[31] Mortazavi A, Williams BA, McCue K, Schaeffer L, Wold B. Mapping and quantifying mammalian transcriptomes by RNA-Seq. Nat Meth 2008;5(7):621–8. http://dx.doi.org/10.1038/nmeth.1226.

[32] Pfeffer U, editor. Cancer genomics - molecular classification, prognosis and response prediction. Netherlands: Springer; 2013.

[33] Pfeffer U, Romeo F, Noonan DM, Albini A. Prediction of breast cancer metastasis by genomic profiling: where do we stand? Clin Exp Metastasis 2009;26(6):547–58.

[34] Eisen MB, Spellman PT, Brown PO, Botstein D. Cluster analysis and display of genome-wide expression patterns. Proc Natl Acad Sci December 8, 1998;95(25):14863–8.

[35] Sørlie T, Perou CM, Tibshirani R, Aas T, Geisler S, Johnsen H, et al. Gene expression patterns of breast carcinomas distinguish tumor subclasses with clinical implications. Proc Natl Acad Sci September 11, 2001;98(19):10869–74.

[36] Dupuy A, Simon RM. Critical review of published microarray studies for cancer outcome and guidelines on statistical analysis and reporting. J Natl Cancer Inst January 17, 2007;99(2):147–57.

[37] van 't Veer LJ, Dai H, van de Vijver MJ, He YD, Hart AAM, Mao M, et al. Gene expression profiling predicts clinical outcome of breast cancer. Nature 2002;415(6871):530–6. http://dx.doi.org/10.1038/415530a.

[38] Paik S, Tang G, Shak S, Kim C, Baker J, Kim W, et al. Gene expression and benefit of chemotherapy in women with node-negative, estrogen receptor–positive breast cancer. J Clin Oncol August 10, 2006;24(23):3726–34.

[39] Albini A, Pennesi G, Donatelli F, Cammarota R, De Flora S, Noonan DM. Cardiotoxicity of anticancer drugs: the need for cardio-oncology and cardio-oncological prevention. J Natl Cancer Inst January 6, 2010;102(1):14–25.

[40] Shi H, Moriceau G, Kong X, Lee M-K, Lee H, Koya RC, et al. Melanoma whole-exome sequencing identifies V600EB-RAF amplification-mediated acquired B-RAF inhibitor resistance. Nat Commun 2012;3:724. http://dx.doi.org/10.1038/ncomms1727.

[41] Turner NC, Reis-Filho JS. Genetic heterogeneity and cancer drug resistance. Lancet Oncol April 2012;13(4):e178–85.

[42] Vitzthum F, Behrens F, Anderson NL, Shaw JH. Proteomics: from basic research to diagnostic application. A review of requirements & needs. J Proteome Res 2005;4(4):1086–97. 2005/08/01.

[43] Ong S-E, Mann M. Mass spectrometry-based proteomics turns quantitative. Nat Chem Biol 2005;1(5):252–62. http://dx.doi.org/10.1038/nchembio736.

[44] Hanke S, Besir H, Oesterhelt D, Mann M. Absolute SILAC for accurate quantitation of proteins in complex mixtures down to the attomole level. J Proteome Res March 2008;7(3):1118–30.

[45] Meissner F, Scheltema RA, Mollenkopf HJ, Mann M. Direct proteomic quantification of the secretome of activated immune cells. Science April 26, 2013;340(6131):475–8.

[46] Rardin MJ, Schilling B, Cheng LY, MacLean BX, Sorenson DJ, Sahu AK, et al. MS1 peptide ion intensity chromatograms in MS2 (SWATH) data independent acquisitions. Improving post acquisition analysis of proteomic experiments. Mol Cell Proteomics May 17, 2015;14(9):2405–19.

[47] Selevsek N, Chang CY, Gillet LC, Navarro P, Bernhardt OM, Reiter L, et al. Reproducible and consistent quantification of the *Saccharomyces cerevisiae* proteome by SWATH-mass spectrometry. Mol Cell Proteomics March 2015;14(3):739–49.

[48] McAlister GC, Nusinow DP, Jedrychowski MP, Wuhr M, Huttlin EL, Erickson BK, et al. MultiNotch MS3 enables accurate, sensitive, and multiplexed detection of differential expression across cancer cell line proteomes. Anal Chem July 15, 2014;86(14):7150–8.

[49] Richards AL, Hebert AS, Ulbrich A, Bailey DJ, Coughlin EE, Westphall MS, et al. One-hour proteome analysis in yeast. Nat Protoc May 2015;10(5):701–14.

[50] Meissner F, Mann M. Quantitative shotgun proteomics: considerations for a high-quality workflow in immunology. Nat Immunol February 2014;15(2):112–7.

[51] Nesvizhskii AI. Proteogenomics: concepts, applications and computational strategies. Nat Methods November 2014;11(11):1114–25.

[52] Olsen JV, Mann M. Status of large-scale analysis of posttranslational modifications by mass spectrometry. Mol Cell Proteomics December 2013;12(12):3444–52.

[53] Yamamoto Y, Kiyoi H, Nakano Y, Suzuki R, Kodera Y, Miyawaki S, et al. Activating mutation of D835 within the activation loop of FLT3 in human hematologic malignancies. Blood April 15, 2001;97(8):2434–9.

[54] Gorre ME, Mohammed M, Ellwood K, Hsu N, Paquette R, Rao PN, et al. Clinical resistance to STI-571 cancer therapy caused by BCR–ABL gene mutation or amplification. Science August 3, 2001;293(5531):876–80.

[55] Bianchi G, Martella R, Ravera S, Marini C, Capitanio S, Orengo A, et al. Fasting induces anti-Warburg effect that increases respiration but reduces ATP-synthesis to promote apoptosis in colon cancer models. Oncotarget March 30, 2015;6(14):11806–19.

[56] Gallien S, Domon B. Detection and quantification of proteins in clinical samples using high resolution mass spectrometry. Methods April 2, 2015;81:15–23.

[57] Edwards NJ, Oberti M, Thangudu RR, Cai S, McGarvey PB, Jacob S, et al. The CPTAC data Portal: a resource for cancer proteomics research. J Proteome Res May 4, 2015;14(6):2707–13.

[58] Oliver SG. Functional genomics: lessons from yeast. Philos Trans R Soc Lond B Biol Sci January 29, 2002;357(1417):17–23.

[59] Pauling L, Robinson AB, Teranishi R, Cary P. Quantitative analysis of urine vapor and breath by gas–liquid partition chromatography. Proc Natl Acad Sci USA October 1971;68(10):2374–6.

[60] Nicholson JK, Wilson ID. Opinion: understanding 'global' systems biology: metabonomics and the continuum of metabolism. Nat Rev Drug Discov August 2003;2(8):668–76.

[61] Fell DA, Wagner A. The small world of metabolism. Nat Biotechnol November 2000; 18(11):1121–2.

[62] Shulaev V. Metabolomics technology and bioinformatics. Brief Bioinforma June 2006; 7(2):128–39.

[63] Ryan D, Robards K. Metabolomics: the greatest omics of them all? Anal Chem December 1, 2006;78(23):7954–8.

[64] Armitage EG, Barbas C. Metabolomics in cancer biomarker discovery: current trends and future perspectives. J Pharm Biomed Anal January 2014;87:1–11.

[65] Dunn WB, Bailey NJ, Johnson HE. Measuring the metabolome: current analytical technologies. Analyst May 2005;130(5):606–25.

[66] Boros LG, Lerner MR, Morgan DL, Taylor SL, Smith BJ, Postier RG, et al. [1,2-13C2]-D-glucose profiles of the serum, liver, pancreas, and DMBA-induced pancreatic tumors of rats. Pancreas November 2005;31(4):337–43.

[67] Wiechert W. 13C metabolic flux analysis. Metab Eng July 2001;3(3):195–206.

[68] Bujak R, Struck-Lewicka W, Markuszewski MJ, Kaliszan R. Metabolomics for laboratory diagnostics. J Pharm Biomed Anal December 25, 2014;113:108–20.

[69] Lin G, Chung YL. Current opportunities and challenges of magnetic resonance spectroscopy, positron emission tomography, and mass spectrometry imaging for mapping cancer metabolism in vivo. BioMed Res Int 2014;2014:625095.

[70] Keren B, Chantot-Bastaraud S, Brioude F, Mach C, Fonteneau E, Azzi S, et al. SNP arrays in Beckwith–Wiedemann syndrome: an improved diagnostic strategy. Eur J Med Genet October 2013;56(10):546–50.

[71] Eckhart AD, Beebe K, Milburn M. Metabolomics as a key integrator for "omic" advancement of personalized medicine and future therapies. Clin Transl Sci June 2012; 5(3):285–8.

[72] Ekins R. Immunoassay standardization. Scand J Clin Lab Invest Suppl 1991;205:33–46.

[73] Spratlin JL, Serkova NJ, Gail Eckhardt S. Clinical applications of metabolomics in oncology: a review. Clin Cancer Res Off J Am Assoc Cancer Res 2009;15(2):431–40.

[74] Shajahan-Haq AN, Cheema MS, Clarke R. Application of metabolomics in drug resistant breast cancer research. Metabolites 2015;5(1):100–18.

[75] Xu XH, Huang Y, Wang G, Chen SD. Metabolomics: a novel approach to identify potential diagnostic biomarkers and pathogenesis in Alzheimer's disease. Neurosci Bull October 2012;28(5):641–8.

[76] Nikolic SB, Sharman JE, Adams MJ, Edwards LM. Metabolomics in hypertension. J Hypertens June 2014;32(6):1159–69.

[77] Kordalewska M, Markuszewski MJ. Metabolomics in cardiovascular diseases. J Pharm Biomed Anal April 25, 2015;113:121–36.

[78] Griffin JL, Wang X, Stanley E. Does our gut microbiome predict cardiovascular risk? A review of the evidence from metabolomics. Circ Cardiovasc Genet February 2015; 8(1):187–91.

[79] Sas KM, Karnovsky A, Michailidis G, Pennathur S. Metabolomics and diabetes: analytical and computational approaches. Diabetes March 2015;64(3):718–32.

[80] Kang J, Zhu L, Lu J, Zhang X. Application of metabolomics in autoimmune diseases: insight into biomarkers and pathology. J Neuroimmunol February 15, 2015;279:25–32.

[81] Kim OY, Lee JH, Sweeney G. Metabolomic profiling as a useful tool for diagnosis and treatment of chronic disease: focus on obesity, diabetes and cardiovascular diseases. Expert Rev Cardiovasc Ther January 2013;11(1):61–8.

[82] Xie G, Zhang S, Zheng X, Jia W. Metabolomics approaches for characterizing metabolic interactions between host and its commensal microbes. Electrophoresis October 2013;34(19):2787–98.

[83] Kinross J, Li JV, Muirhead LJ, Nicholson J. Nutritional modulation of the metabonome: applications of metabolic phenotyping in translational nutritional research. Curr Opin Gastroenterol March 2014;30(2):196–207.

[84] Freemark M. Metabolomics in nutrition research: biomarkers predicting mortality in children with severe acute malnutrition. Food Nutr Bull March 2015;36(1 Suppl.):S88–92.

[85] Wilcock A, Begley P, Stevens A, Whatmore A, Victor S. The metabolomics of necrotising enterocolitis in preterm babies: an exploratory study. J Matern Fetal Neonatal Med Off J Eur Assoc Perinat Med Fed Asia Ocean Perinat Soc Int Soc Perinat Obstet March 19, 2015:1–5.

[86] Zhang J, Carey V, Gentleman R. An extensible application for assembling annotation for genomic data. Bioinformatics January 2003;19(1):155–6.

[87] Cox J, Mann M. MaxQuant enables high peptide identification rates, individualized p.p.b.-range mass accuracies and proteome-wide protein quantification. Nat Biotechnol December 2008;26(12):1367–72.

[88] Cox J, Mann M. 1-D and 2-D annotation enrichment: a statistical method integrating quantitative proteomics with complementary high-throughput data. BMC Bioinforma 2012;13(Suppl. 16):S12.

[89] Chen EY, Tan CM, Kou Y, Duan Q, Wang Z, Meirelles GV, et al. Enrichr: interactive and collaborative HTML5 gene list enrichment analysis tool. BMC Bioinforma 2013;14:128.

[90] Huttenhower C, Haley EM, Hibbs MA, Dumeaux V, Barrett DR, Coller HA, et al. Exploring the human genome with functional maps. Genome Res June 2009;19(6):1093–106.

[91] Galperin MY, Rigden DJ, Fernandez-Suarez XM. The 2015 nucleic acids research database issue and molecular biology database collection. Nucleic Acids Res January 2015;43(Database issue):D1–5.

[92] Nannya Y, Sanada M, Nakazaki K, Hosoya N, Wang L, Hangaishi A, et al. A robust algorithm for copy number detection using high-density oligonucleotide single nucleotide polymorphism genotyping arrays. Cancer Res July 15, 2005;65(14):6071–9.

[93] Lohr JG, Stojanov P, Lawrence MS, Auclair D, Chapuy B, Sougnez C, et al. Discovery and prioritization of somatic mutations in diffuse large B cell lymphoma (DLBCL) by whole-exome sequencing. Proc Natl Acad Sci USA March 6, 2012;109(10):3879–84.

[94] Mallick P, Kuster B. Proteomics: a pragmatic perspective. Nat Biotech 2010;28(7): 695–709. http://dx.doi.org/10.1038/nbt.1658.

Regulatory Process in the United States of America, Europe, China, and Japan

Section 1: Regulatory Process in the United States of America

Elizabeth K. Leffel[1], Ross D. LeClaire[2]
[1]Leffel Consulting Group, LLC, Berryville, VA, USA; [2]The Translational Bridge, LLC, Albuquerque, NM, USA

INTRODUCTION AND BACKGROUND

A discussion of translational medicine must include the incorporation of the regulatory strategy for a "medicinal product" development program. Medicinal products include drugs, biologics, vaccines, and devices. Each is unique and has individual regulatory requirements. For the purposes of this chapter, the focus will be on drugs in order to introduce concepts in a concise manner. The path from basic research that produces vital discoveries to the result of a licensed medical product that benefits patients can be a formidable undertaking [1]. Bringing a new drug to market typically takes more than a decade, $2 billion dollars, and has a failure rate in excess of 95% [2,3]. The concept of translational research continues to evolve. It encompasses basic drug discovery and applied preclinical research; the characterization of manufacturing conditions and analytical assays; evaluation of preclinical outcomes as they apply to approval for human use; and finally the assessment of human health outcomes.

The National Center for Advancing Translational Science of the National Institutes of Health identifies the regulatory process as one of the major challenges of translational medicine. The regulatory approval process in the United States (US) is governed by the Food and Drug Administration (FDA). The FDA is the US government agency with a total of nine centers (or offices). The largest is the Center for Drug Evaluation and Research (CDER), which regulates "over-the-counter and prescription drugs, including biological therapeutics and generic drugs" [4]. A second is the Center for Biologics Evaluation and Research (CBER), which also regulates biological products for human use [5]. Sponsors must be aware of which FDA divisions are reviewing drug applications. ("sponsor" is the formal reference to the company or entity ultimately submitting the drug application.) The FDA encourages sponsors to contact the FDA early in the development process to identify the appropriate pathways to be navigated; "guidance documents" are published to assist sponsors in selecting appropriate meetings and outlining the process by which to secure those meetings [6].

This chapter will describe the drug approval process that is governed by the FDA by integrating descriptions of the drug development stages into the regulatory milestones. Throughout the process, key considerations are highlighted in which translational medicine is critical for completing ethical, scientifically sound, and efficient research for submission to the FDA for drug license approval. The diverse requirements of translational research are beyond individual disciplines and necessitate the formation of interdisciplinary collaborations [7].

The first major regulatory milestone will likely be a pre-IND (Investigational New Drug) meeting (to be discussed in detail in sections below). Before that occurs, a company will have completed an extensive amount of research and development work on the target drug, in addition to a large body of preclinical studies. The stages of drug development that occur prior to a pre-IND meeting would be the "drug discovery" and "preclinical" stages. The second major regulatory milestone is the submission of the IND application. There are years of research, development, manufacturing optimization, and preclinical testing that must be done to generate enough data to complete an IND package. When a company has successfully received an IND approval, clinical (human) trials may begin, thus the program enters the "clinical stage" of drug development. On occasion, there are ongoing preclinical testing and biomarker analyses in parallel with the clinical trials to provide translational data that will ultimately be submitted to the FDA. The final, major regulatory milestone for a sponsor to obtain a license to sell a drug is to submit a new drug application (NDA). Some "drugs" are produced from biologic materials, so instead of applying for an NDA, new biologic drugs are approved using a biologics license application (BLA). The submission for either an NDA or BLA will each require data to support the drug's safety and effectiveness and demonstrate that the drug can be manufactured safely.

DRUG DEVELOPMENT: DISCOVERY STAGE

Prior to a drug target being identified, the pathogenesis of a disease must be elucidated, which may take decades. Once a "target" is identified, the drug discovery stage begins as screenings that identify the correct molecules that will interact with the target to produce the desired effect—preventing, curing, or treating a disease. This potential candidate drug must then be produced at a "bench scale," which means small quantities are manufactured in the laboratory for use in in vitro and in vivo assessment. Research at this stage is not directly subject to strict regulatory oversight, as described by Kang et al., where the impact of the regulatory process on research and development is discussed in detail. The data may be presented to the FDA in a pre-IND meeting supporting ethical study designs for larger preclinical studies.

DRUG DEVELOPMENT: PRECLINICAL STAGE

The next stage of drug development is generally referred to as "preclinical." As the name implies, this body of work must be completed prior to the FDA allowing a clinical trial to begin in humans. Preclinical testing is completed in both in vitro and in vivo models to determine if the drug is safe and efficacious. These studies evaluate a drug's safety, efficacy, and potential toxicity in vitro, in silico, and in vivo (animal models) using a wide range of drug concentrations, dosing regimens, and drug formulation(s). It is during this stage that information is gathered to ultimately effect the "translational" end points for a program. Animal models must be carefully considered and chosen based on the relevance to the disease manifestation in humans. The first safety signals are gathered in a series of many toxicology studies. Sponsors must begin to identify a panel of potential biomarkers and investigate how to adequately measure them both in animal models (perhaps even in cell cultures) and in humans. Because it is important to understand the biological activity of the drug, a preclinical research program also often includes pharmacodynamic (PD) and pharmacokinetic (PK) studies in a relevant animal model to predict therapeutic doses in humans, stressing the importance of the ability to adequately bridge preclinical and clinical stages.

Depending on the drug class and mechanism of action, FDA may request other specific studies. For example, if the drug (nonbiologic) is a small molecule, drug metabolism will be evaluated prior to beginning a clinical trial; "absorption, distribution, metabolism, and excretion" (ADME) studies are generally completed in healthy animals. Another example of a specific investigation required is found in the case of vaccines. Immunology studies need to be completed to determine correlates of immunity or in contrast, evidence of reactogenicity.

Not only will the preclinical data inform the design of the first clinical trials, but also a critical component that shapes the manufacturing process. In parallel to the preclinical studies, manufacturing processes, analytical assays, and product release specifications are being developed. As the toxicological and bioavailability profile of the investigational drug unfolds, it may necessitate a change in formulation or production in order to reduce toxicity or increase absorption. The preclinical stage culminates in submitting the IND application to the FDA.

There are several key "regulatory milestones" along the drug development timeline. In addition, there are significant opportunities for meeting with the FDA, and these are discussed in the text below as they fall in the regulatory timeline. The meetings are described in greater detail in an FDA guidance for industry [6].

Regulatory Milestone: Pre-IND Meeting

Prior to submitting the IND package, a sponsor may request a pre-IND Type B meeting (21 CFR (Code of Federal Regulations) 312.82 (a)). The primary purpose of this meeting is to review the preclinical data to receive FDA concurrence that it is sufficient to allow initiation of clinical trials. In addition, it is an opportunity to discuss the design and draft protocol for the Phase 1 trial(s). These meetings are critical if an investigational drug has a novel indication, is a new molecular entity, or has generated any questionable data in toxicology or PK studies.

There are some very specialized regulatory options at the FDA (e.g., orphan drugs). One such special rule was written to address cases in which it is unethical or infeasible to test the drug in people (e.g., anthrax vaccine or treatment for pneumonic plague); the FDA Animal Rule (21 CFR 314.600 for drugs; CFR 601.90 for biologics) [8,9] applies to these programs. The FDA may grant marketing approval based on adequate and well-controlled animal efficacy studies. Pre-IND meetings become extremely critical to the successful licensure of such a drug because the FDA must make decisions without the benefit of efficacy data in humans. Frequently a sponsor developing a drug via this pathway will have multiple pre-IND meeting to ensure that the preclinical and nonclinical data will successfully translate into predictable human benefit.

Regulatory Milestone: IND Submission

The CFR contains the laws by which the FDA is governed during the review and granting of the IND. In most cases, there is a specific regulation (Chapter 21 of the CFR) written to apply for each type of product that might be developed. The FDA also writes and publishes "guidance documents," which contain useful information on requirements [10,11]. There are two IND categories—commercial and research (noncommercial) and three IND types are as follows: an *Investigator IND* submitted by a physician who is responsible for conducting the trial(s) and overseeing the administration of the investigational drug; an *Emergency Use IND* gives the FDA the authority to allow the administration of an investigational drug to humans in an emergency situation when there is no sufficient time for submission of an IND in accordance with federal regulations; and a *Treatment IND* which can be submitted for experimental drugs for life-threatening diseases to be used while the final clinical data are in review with the FDA.

During the IND process, the FDA will review all preclinical data for the toxicology and animal pharmacology studies in addition to manufacturing information and clinical protocols with plans for human testing. The clinical protocol must include details that will explain how the sponsor will address the following: selection criteria: study group sizes, duration of the study, description of control groups, methods to reduce research bias, drug administration route, dosage regimen, end points, and finally, the data collection, review, and analysis methods. The primary objective of the IND review is for the FDA to determine that the investigational drug is safe, has demonstrated the capability to benefit humans, and there is an appropriate clinical plan to safely evaluate the new drug.

Once an IND is filed, the FDA has 30 days to review the packet. If the FDA has not contacted the sponsor to place a clinical hold on the investigational drug, then the trial may begin on day 31. Of the 4 phases of clinical trials, 1–3 must be completed prior to an application for a license to market the drug. The trials begin with small groups of healthy volunteers and expand into larger trials with groups of the target population either at risk or affected by the disease.

DRUG DEVELOPMENT: CLINICAL TRIALS STAGE

At any point from IND onward, the FDA can impose a clinical hold, i.e., prohibit the trial(s) from proceeding, or stop a trial that has started for reasons of safety concerns, discovery of a sponsor's failure to disclose known risk, or any other reason allowed by law.

In some cases, a sponsor will elect to conduct a trial that is referred to as "Phase 0"; these are first-in-human clinical trials that occur prior to Phase 1. The purpose of a Phase 0 is to streamline the drug approval process. These are usually PD and/or PK trials administering a single dose of the drug. Each study group of healthy human volunteers will sequentially receive a higher single dose than the previous group. The trial is designed to document the ADME profile of the drug (or in some cases, the biologic) to confirm that the mechanism of action results in predicted biological activity. Phase 0 studies are, in a sense, the first key to the final step in the translational medicine pathway as they are used to confirm that the drug behaves as expected based on preclinical outcomes. If the results of the Phase 0 trial are not as predicted, then the deviations can be evaluated and addressed early, thus avoiding human risk and a drug development delay incurring subsequent additional development expense.

Drug Development: Clinical Trial Stage—Phase 1

The primary purpose of Phase 1 clinical trials is to evaluate product safety and dosage with drug manufactured by a process reflective of the final procedures under current good manufacturing practices (cGMP) [12]. The total group size is small (20–80 healthy volunteers or people with the disease/condition) [13]. Smaller cohorts are randomly identified to evaluate safety, method of administration, safe dosage ranges, and potential side effects. A drug's side effects could be subtle or long term or erratic. As such, Phase 1 trials are not expected to identify all side effects. If the product is deemed safe, then drug development moves into a Phase 2 trial; approximately 70% of drugs tested in a Phase 1 trial will advance into Phase 2 [14].

Biological markers (i.e., biomarkers) are established in the preclinical phase and then evaluated in Phase 1 clinical trials. Biomarkers are required in translational medicine to link PD/PK/ADME data and functional responses between preclinical and clinical work in order to get the first indications of the efficacious dose for the investigational drug. Biomarkers are often used as a fundamental component of product development [15], and outcomes may serve as the basis of a go/no-go decision-making tool. A biomarker must possess a characteristic(s) that can be measured consistently as an indicator of a normal biologic process, a pathogenic process, and finally a pharmacologic response to a therapeutic intervention. Bioanalytical assay development should be completed by this stage and validated assays (21 CFR 58) should be in use. For a biomarker to be useful in drug development decision making, it must be an accurate and sensitive indicator of clinical efficacy.

Regulatory Milestone: End-of-Phase 1 Meeting

At the conclusion of the Phase 1 trials, a sponsor may request an end-of-Phase 1 meeting with the FDA (21 CFR 312.82). This type of meeting is generally requested for a drug designed to treat a severely debilitating or life-threatening disease. The purpose of this Type B meeting is to review results of the Phase 1 study, present the Phase 2 protocol, and reach agreement on plans for the Phase 2 program.

Drug Development: Clinical Trial Stage—Phase 2

The Phase 2 trials allow for the estimation of safety in a larger (100–300) population of patients that have the disease or condition. Efficacy end points in Phase 2 trials are often compared to placebo control groups, but the trials are not designed to solely determine the efficacy of the drug because the drug is given to a limited number of patients under a well-controlled condition. These trials will gather more data to better define the profile for adverse events or placebo effects. An *adverse event* is defined by the FDA as an "undesirable experience associated with the use of a medical product in a patient" and is reportable if the "outcome results in death, life-threatening event, hospitalization, disability/permanent damage, congenital anomaly/birth defect, requires medical intervention or other serious medical event." A *placebo effect* is one in which a participant perceives or reports nonspecific benefit from receiving any treatment; the person is unaware that the treatment is not the drug being tested. The placebo effect is controlled by inclusion of randomly assigned patients who receive only a placebo, and both the participant and personnel conducting the trial are blind to the treatment group (i.e., double-blinded trial). Ethical issues may be present with the assignment of a patient to a placebo group because it breaches their right to receive the best available treatment. The identification and resolution of ethical issues is guided by the Declaration of Helsinki (see http://www.who.int/bulletin/archives/79(4)373.pdf).

In parallel with Phase 2, often other trials are performed, which include expected drug interactions and more specific safety studies where needed. It is estimated that only 33% of drugs move from Phase 2 into Phase 3 trials [14]. When a successful Phase 2 trial is completed, a sponsor may request an end-of-Phase 2 meeting prior to entering the Phase 3 trial.

Regulatory Milestone: End-of-Phase 2 Meeting

The primary purpose of this Type B (21 CFR 312.47 (b) (1)) meeting is to determine if it is safe to proceed to a Phase 3 trial. The sponsor seeks agreement on Phase 3 study designs, and safety and efficacy end points. In addition, it is an opportunity to update the IND on manufacturing processes and PD/PK data while assuring the FDA that all translational preclinical data will support licensure. At this meeting, it would be appropriate to discuss the need for any additional information that may support a license, such as pediatric requirements.

Drug Development: Clinical Trial Stage—Phase 3

These trials are the final confirmation of product safety and efficacy before the sponsor files an NDA or BLA. Group sizes are larger (300–3000 people with the disease or condition) and justified statistically to indicate a product is efficacious, the risk of side effects are reasonable compared to the benefit from receiving the drug, and there are no serious adverse events (or that they are manageable or balanced against benefit). In some cases, the FDA will require a sponsor to determine if the treatment is better than standard of care (i.e., prove "treatment benefit") [16,17a]. Participants are randomly assigned to the control group (standard-of-care treatment) or study group (test treatment) using a computerized process to avoid bias in the clinical trial.

Regulatory Milestone: Pre-NDA/Pre-BLA Meeting

Requesting a pre-NDA/BLA meeting (21 CFR 312.47 (b) (2)) will allow the sponsor to review the final package material with the FDA team. The primary purpose of this Type B meeting is to determine the adequacy of the application, and if there is a need for an advisory committee meeting instead of the standard review or if plans for risk evaluation and mitigation strategies are needed.

Regulatory Milestone: NDA or BLA Submission

The Food, Drug, and Cosmetic Act, of 1938 required new drug applicants, for medical products intended for commercial use in the US, to submit an NDA. A sponsor will file an NDA when the investigational drug is not biologically derived. The Act mandated a premarket review of the product's safety (the efficacy requirement was added by the 1962 Kefauver-Harris Amendments to the Act), label requirements and gave the FDA authority to inspect/audit the manufacturing facilities. The mandated times for review are defined in the Prescription Drug User Fee Act. In cases of extreme situations, such as the recent Ebola epidemic in West Africa, the FDA has instituted expedited processes and informative updates are recorded on the Web site [17b]. This has allowed the FDA to respond very quickly so that products enter testing phases during the period of dire need.

The data gathered throughout all stages of drug development may become part of the NDA and the quantity of information submitted can vary significantly as they depend on the nature of the product, precedence of the mechanism of action, and information at the time of submission. The documentation required in an NDA must adequately describe the drug, including results of the following: chemistry, manufacturing, and controls; preclinical/nonclinical studies (PK/PD, toxicology, efficacy, etc.); and clinical trials. The objective of compiling all of these data is so the submission provides enough information for the FDA to determine the following points: (1) the drug is safe and effective for its proposed use(s), (2) the benefits of use outweigh the risks, (3) the proposed labeling is appropriate, and (4) the manufacturing methods and quality controls used are adequate to preserve the products identity, potency, quality, and purity. Current guidance on the NDA application may be found on the FDA Web site [17c]. Upon FDA approval of an NDA, the drug can be legally marketed starting that day in the US.

The FDA has 60 days to conduct a preliminary assessment of the NDA to determine if it is adequate to permit a functional review. If incomplete, the FDA will reject the application and issue a "refuse-to-file" letter (21 CFR 314.101) [18] detailing where the application has failed to meet requirements. A sponsor can request an informal meeting, called a filing conference, if the FDA refuses to file an application. After the meeting, the sponsor can request that the FDA file the application and the agency will do so, over protest (21 CFR 314.101), and review it as filed, or the sponsor can amend the application and resubmit it. If the NDA is deemed complete, the submission is assigned an NDA tracking number and will then undergo either a standard review (approximately 10–12 months) or expedited review (approximately 6 months).

Prior to taking any decision on the NDA, there will be a (1) preliminary review to validate the raw data submission; (2) secondary review of the results of the primary review by the division director to determine if a recommendation to approve or disapprove is warranted; and (3) final review for determination of approval or issuance of a "complete response" letter (21 CFR 314.110) [19].

In addition, advisory committee [20] meetings are mandated for certain products or if there are critical questions on the safety and efficacy of the product. Although the FDA is not obligated to follow the advice or recommendations from the advisory committee, it frequently does. Following the review of data submitted in support of safety and efficacy, FDA may carry out preapproval inspections (PAIs) that may include the manufacturing facilities as well as facilities or sites where the drug was tested in nonclinical and/or clinical studies. The decision to conduct a PAI is at the discretion of the FDA. Triggers are likely to be an application for a new molecular entity, a new therapeutic biological product, a priority NDA, the first NDA filed by the sponsor, a "for-cause" inspection, or if the cGMP status is unknown (e.g., more than 2 years since the facility was inspected). The FDA inspects foreign and domestic facilities that will manufacture the drug for the US market or that have completed supporting studies for approval of the product in the US. The results of the PAI inspection allow the agency to determine the accuracy and credibility of the data submitted. Finally, the FDA reviews product labeling to determine if it correctly reflects the product's safety and if it allows physicians/patients to compare benefits and risks.

A sponsor will file a BLA, instead of an NDA, when the investigational "drug" is biologic. BLAs are generally reviewed by CBER [21] and are regulated under 21 CFR 600–680. The area of biologics encompasses products that fall into the following categories: blood products, tissue products, and vaccines. Due to the nuances in each area, the FDA has published specific "rules" (found in the CFR) for each because the development pathway is a bit different depending on the product. FDA guidances should be consulted and pre-IND meetings should be scheduled appropriately to ensure that a sponsor is interacting with the correct review division at the FDA and complying with the appropriate rules.

Pertinent manufacturing, safety information, and instructions for the use of the drug are provided in an FDA-approved "product label" (21 CFR 201.5). The main purpose of a drug label (i.e., package insert) is to provide health-care providers and patients with adequate information and directions for the safe use of the drug. The product label should be drafted by the IND submission stage and updated as needed, up to the time of filing the NDA.

COMPLETE RESPONSE LETTER

In 2009, the FDA amended its regulations on NDAs. It discontinued the use of "approvable letters" and "not approvable letters." If an NDA was not approved, the sponsor would receive a complete response letter to indicate that the review cycle for an application was complete but it was not ready for approval (21 CFR 314.110; 314.125; 314.127). Upon receiving a complete response letter, a sponsor can choose to (1) resubmit the application and address all deficiencies outlined in the letter, (2) withdraw the application, or (3) request a hearing with the FDA to provide the sponsor an opportunity to clarify the basis for denying approval.

END OF REVIEW CONFERENCE

After the FDA has issued a complete response letter, the sponsor may request a Type C meeting (21 CFR 314.102) with the review team. The objective of these types of meetings is to discuss what the sponsor must do to strengthen the NDA package in order for the application to be approved.

Regulatory Milestone: Postmarketing Surveillance

Active postmarketing surveillance for adverse events of medical products is considered essential. All possible side effects of a product may not be identified during the preapproval process because it involves a limited number of subjects with limited population diversity. The FDA maintains a database of postmarketing surveillance and risk assessment to identify adverse events that did not appear during the preapproval process called the Adverse Event Reporting System. Adverse events and errors are coded using medical dictionary for regulatory activities terminology and adhere to the international safety reporting guidance issued by the International Conference on Harmonization [22]. The US also utilizes a "MedWatch" system, which is the program used for health professionals and the public to report serious adverse events related to medical products. In addition, during the postmarketing surveillance, the FDA will continue to verify adherence to the conditions of approval described in the application and that the drug is manufactured in a consistent and controlled manner. This is done by periodic impromptu inspections of drug production and control facilities. Manufacturers must also submit error, accident, and drug quality reports when deviations from cGMP regulations occur.

In the US, the FDA has the authority to approve medicinal products for marketing and sale. The "drug development" process is lengthy, costly, and complicated. At every stage of the regulatory process, translational science plays a role by building on basic research and integrating data from in vitro and animal models with that obtained from humans, to result in a drug that benefits people.

LIST OF ACRONYMS AND ABBREVIATIONS

ADME Absorption, distribution, metabolism, and excretion
BLA Biologics license application
cGMP Current good manufacturing practices
CBER Center for Biologics Evaluation and Research
CDER Center for Drug Evaluation and Research
CFR Code of Federal Regulations
FDA Food and Drug Administration
IND Investigational New Drug
NDA New drug application
PAIs Preapproval inspections
PD Pharmacodynamic
PK Pharmacokinetic
US United States

References

[1] Drolet BC, et al. Translational research: understanding the continuum from bench to bedside. Transl Res 2010;157(1):1–5.
[2] Keating L. Government support of translation science: a promising way to bridge the development gap and increase technology commercialization to support the American economy. J High Technol Law 2013;650:651–78.

[3] National Institutes of Health website. About NCATS. [Online]. Available: http://www.ncats.nih.gov/about/about.html.

[4] FDA. About the center for drug evaluation and research. [Online]. Available: http://www.fda.gov/AboutFDA/CentersOffices/OfficeofMedicalProductsandTobacco/CDER; 2014.

[5] FDA. USFDA. [Online]. Available: http://www.fda.gov/AboutFDA/CentersOffices/OfficeofMedicalProductsandTobacco/CBER/default.htm; 2015.

[6] FDA. Guidance for industry formal meetings between the FDA and sponsors or applicants. 2009.

[7] Kong HH, et al. Bridging the translational research gap. J Invest Dermatol 2010;130:1478–80.

[8] FDA. FDA news release. [Online]. Available: http://www.fda.gov/NewsEvents/Newsroom/PressAnnouncements/ucm302220.htm; 2012.

[9] FDA. sBLA approval. [Online]. Available: http://www.accessdata.fda.gov/drugsatfda_docs/appletter/2015/103353Orig1s5183ltr.pdf; 2015.

[10] FDA. Guidance for clinical investigators, sponsors, and IRBs. Investigational new drug applications (INDs) – determining whether human research studies can be conducted without an IND. 2013.

[11] FDA. Investigational new drug (IND) application. [Online]. Available: http://www.fda.gov/drugs/developmentapprovalprocess/howdrugsaredevelopedandapproved/approvalapplications/investigationalnewdrugindapplication/default.htm; 2014.

[12] FDA. Guidance for industry cGMP for phase 1 investigational drugs. 2008.

[13] FDA. The FDA's drug review process: ensuring drugs are safe and effective. [Online]. Available: http://www.fda.gov/drugs/resourcesforyou/consumers/ucm143534.htm; 2014.

[14] FDA. Step 3: clinical research. [Online]. Available: http://www.fda.gov/forpatients/approvals/drugs/ucm405622.htm#phases; 2015.

[15] Park JW, et al. Rationale for biomarkers and surrogate end points in mechanism-driven oncology drug development. Clin Cancer Res 2004;10(11):3885–96.

[16] Migone TS, et al. Added benefit of raxibacumab to antibiotic treatment of inhalational anthrax. Antimicrob Agents Chemother 2015;59(2):1145–51.

[17] [a] Crobu D, et al. Preclinical and clinical phase I studies of a new recombinant Filgrastim (BK0023) in comparison with Neupogen. BMC Pharmacol Toxicol 2014;15.
[b] FDA. Ebola response updates from FDA. [Online]. Available: http://www.fda.gov/EmergencyPreparedness/Counterterrorism/MedicalCountermeasures/ucm410308.htm.
[c] FDA. Development & approval process (drugs). [Online]. Available: http://www.fda.gov/Drugs/DevelopmentApprovalProcess/default.htm.

[18] FDA. Good review practice: refuse to file. Office of new drugs, manual of polices and procedures. 2013.

[19] FDA. Complete response letter final rule. [Online]. Available: http://www.fda.gov/Drugs/GuidanceComplianceRegulatoryInformation/LawsActsandRules/ucm084138.htm; 2009.

[20] FDA. Advisory committees. [Online]. Available: http://www.fda.gov/AdvisoryCommittees/default.htm; 2015.

[21] FDA. USFDA. [Online]. Available: http://www.fda.gov/BiologicsBloodVaccines/DevelopmentApprovalProcess/BiologicsLicenseApplicationsBLAProcess/; 2014.

[22] International Conference on Harmonization. Maintenance of the ICH guideline on clinical safety data management: data elements for transmission of individual case safety reports E2B(R2); 2001.

Section 2: Regulatory Process in Europe, China, and Japan

Parviz Ghahramani

Chief Executive Officer, Inncelerex, Jersey City, NJ, USA; Affiliate Professor, School of Pharmacy, University of Maryland, Baltimore, USA

Parviz.Ghahramani@inncelerex.com

INTRODUCTION AND BACKGROUND

Medicinal products include drugs, biologics, vaccines, and devices. Regulations governing these products have many differences in nature; where relevant, the differences and similarities are highlighted in this chapter.

The cost of development to get a drug approved has increased from $800M to $2.6B per new approved drug from 2003 to 2014 [1]. A major part of this increase is attributed to increasing regulatory requirements for more complex, longer, and larger clinical studies and also attributed to increased attrition rate of compounds during the development, which all drive the R&D costs higher. Translational research has the potential to reduce these increasing trends by aiding more appropriate compound selection at early stages of development, which would in turn reduce the time and cost of clinical studies and achieving drug approvals at a lower R&D cost.

This chapter provides an overview of drug regulatory processes related to European Union (EU), which is one of the two largest markets for medicinal products (the US is the other). The last part of this chapter will also briefly provide some key features of the regulatory authorities in Japan and China.

This chapter will provide an overview of the drug approval process that is governed by regulatory authorities. Where relevant, this chapter will discuss impacts that translational medicine can make to reduce time and cost of research for drug applications and approvals of drug licenses.

DRUG REGULATORY APPROVAL PROCESS IN EUROPE

As of 15 April 2015, there are 28 member states in the EU. European Medicines Agency (EMA) is responsible for regulatory assessment and recommendations for approval of new drugs. This process would result in a marketing authorization (MA) that is valid throughout EU as well as Iceland, Norway, and Liechtenstein.

Overall, there are two regulatory stages for a drug approval for human use. The first stage is clinical trial application (CTA) and the second is MA application (MAA). CTA is done at individual national member state level, which authorizes conduct of trials in humans in the member state for a compound that is in investigational stage. MAA is the stage of the process dealing with evaluation of the investigational drug for marketing approval. There are different options for MAA that will be discussed in this chapter. The two stages described above are analogous to the Investigational New Drug (IND) and new drug application (NDA) in the US, but there are many differences between the EU and the US in the regulatory requirements and processes.

There are generally three options for MAAs: (1) centralized, (2) decentralized, and (3) mutual recognition procedure [2]. These are described below in more details.

Centralized MAA Procedure

An applicant would obtain an MA that is accepted throughout the EU member states, Iceland, Norway, and Liechtenstein. The centralized process came into effect in 1995. This process is mandatory for the following products:

- medicinal products manufactured using biotechnological processes
- orphan medicinal products
- products not authorized in the EU before 20 May 2004 only if they are intended for the treatment of AIDS, cancer, neurodegenerative disorder, or diabetes

For centralized procedure, the applicant sends an application directly to the EMA that is then evaluated by the Committee for Medicinal Products for Human Use (CHMP) [3]. The process results in a CHMP recommendation to European Commission (EC) who ultimately decides on approval of a drug. The EC approval is valid throughout the member states. The membership and structure of the CHMP is described later in this chapter.

There is also the Committee on Herbal Medicinal Products (HMPC) that is responsible for creation and maintenance of a list of herbal substances, preparations, and combinations thereof for human use. The processes and responsibilities of HMPC are beyond the scope of this chapter and will not be discussed.

When an MAA is submitted, the EMA scientific committee appoints a rapporteur and a co-rapporteur who would coordinate the assessment of the EMA and prepare the draft reports. External experts may be called to review and provide input to the draft report, which is then submitted to the CHMP. The comments or objections of the CHMP are then communicated to the applicant. Replies from the applicant are communicated to CHMP for preparation of the final assessment report. Throughout the process, rapporteur and co-rapporteur remain the intermediary between applicant, CHMP, and external experts for communications.

Once the evaluation is completed, the CHMP provides a favorable or unfavorable opinion to EC in order to grant or reject the MA sought by the applicant. In the case of a favorable opinion, the applicant would also receive a draft summary of the product's characteristics (SPC), the package leaflet, and the texts proposed for the various packaging materials translated into all EU official languages (Figure 1).

Flow chart (Figure 1) shows the sequence and time lines of important milestones in the centralized procedure for MAA. The time limit for the evaluation through centralized procedure is 210 days. But, notice that the clock may stop at certain milestones.

Products of significant interest in terms of public health can request an accelerated assessment. In that case, a decision could be reached in 150 days, instead of 210 days, if no outstanding issues remain at day 120. Success for accelerated approval is rare though, and application for an accelerated approval must meet a set of defined criteria.

For accelerated approvals, the centralized procedure allows the registration of products with less data than required for other normal approvals. This is done only under exceptional circumstances if the product is promising and where the drug is for a serious disease with unmet medical needs. The data and evidence for full efficacy and safety profile have to be

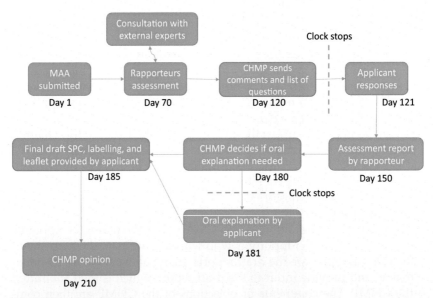

FIGURE 1   Flow chart and timelines of key milestones for centralized marketing authorization application (MAA) procedure. CHMP, Committee for Medicinal Products for Human Use; SPC, summary of the product's characteristics.

demonstrated through further studies even after accelerated approvals. The status of the MA must be confirmed annually until a full MA can be granted.

Mutual Recognition Procedure

To be eligible for this procedure, a medicinal product must have already received an MA from at least one member state. This procedure has been mandatory since 1 January 1998. Any MA granted by one or more EU member state's national authority can be used to apply for mutual recognition by other member state's national authorities. Mutual recognition procedure (MRP) can be requested from one or more member states other than the original member state where there is prior approval [2].

An identical MAA is submitted to all member states that the applicant selects to address. As soon as one member state decides to assess the product that member state becomes the reference member state (RMS). The RMS must notify other member states, which then become concerned member states (CMS). The CMS would then suspend their own evaluations and await the decision for approval (or nonapproval) by the RMS. The assessment by RMS may take up to 210 days after which it would result in granting MA in the RMS. In some cases, an MA may have been granted by the RMS. In such a case, the RMS would update the existing assessment report within 90 days. As soon as the assessment by RMS is concluded, reports are sent to all CMS, along with the SPC, labeling, and package leaflet. Each CMS then has 90 days to recognize (or deny) the decision of the RMS, the SPC, labeling, and package leaflet as approved. National MAs are granted within 30 days after acknowledgment of the agreement.

Should any CMS deny to recognize the original national authorization granted by the RMS, on the grounds of potential serious risk to public health, the issue will be referred to the Coordination Group for Mutual Recognition and Decentralized Procedures (CMD, see more details later about CMD in this chapter) that would make all efforts to reach an agreement within 60 days among all member states addressed by the application. If no agreement is reached, the case is referred to EMA scientific committee (CHMP), for arbitration. The decision of CHMP is then forwarded to the EC. Figure 2 shows the flow chart for the MRP and its key milestones.

Decentralized MAA Procedure

This procedure is based on recognition by other EU member national authorities of a first assessment performed by one member state. This procedure is very similar to MRP; however, it can be applied only for medicinal products that have not received an MA at the time of application. An identical MAA is submitted simultaneously to several competent authorities of

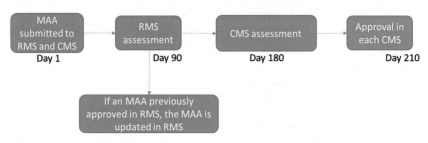

FIGURE 2   Flow chart and time lines of key milestones for mutual recognition procedure. MAA, marketing authorization application; RMS, reference member state; CMS, concerned member states.

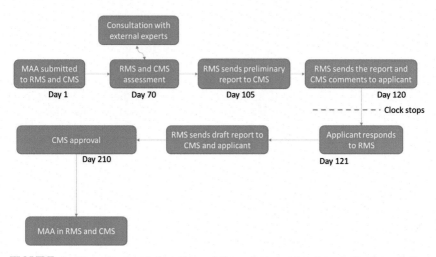

FIGURE 3   Flow chart and time lines of key milestones for decentralized marketing authorization application (MAA) procedure. RMS, reference member state; CMS, concerned member states.

member states that the applicant intends to market the product. One member state is selected by the applicant as the RMS and others would become the CMS. At the end of the procedure, the draft assessment report, SPC, labeling, and package leaflet, as proposed by the RMS, are approved by CMS. Figure 3 shows the flow chart of the decentralized procedure and the key timelines.

OVERVIEW OF VARIOUS COMMITTEES IN EMA

There are several committees within the EMA that deal with specific assessments or evaluations. The following describes the role and the construct of a few committees within the EMA that are concerned with human medicinal products.

Committee for Medicinal Products for Human Use

The members of CHMP are assigned by EU member states in conjunction with EMA's management board. The members serve on the committee for a period of 3 years and their appointment is renewable.

CHMP has a chair who is elected by active CHMP members. Each member state nominates one member and an alternate for the CHMP committee. Also, one member and one alternate are nominated by Iceland and by Norway. Up to five other members can be appointed among the experts nominated by the member states or by EMA. These five additional CHMP members are selected based on their expertise to provide insight for particular scientific areas, when needed.

Pharmacovigilance Risk Assessment Committee

Pharmacovigilance Risk Assessment Committee (PRAC) is responsible for assessing and monitoring safety issues for human medicines. PRAC provides the safety recommendations to CHMP. PRAC is composed of the following members:

- a chair and a vice chair who are appointed by active PRAC members
- one member and an alternate nominated by each of the 28 EU member states, by Iceland, and by Norway
- six independent scientific experts nominated by the EC
- EC (with consultation of European Parliament) nominates one member and an alternate to represent health care professionals; similarly EC nominates one member and an alternate to represent patients organizations

Committee for Orphan Medicinal Products

Committee for Orphan Medicinal Product (COMP) (instead of CHMP) is responsible for review of applications seeking orphan medicinal product designation. Orphan medicinal products are defined as products that are intended for diagnosis, prevention, or treatment of life-threatening diseases or very serious conditions that affect no more than 5 in 10,000 people in the EU. The COMP is composed of the following members: one chair elected by the active COMP members; one member nominated by each of the 28 member states, one member nominated by Norway, and one by Iceland; three members nominated by EC to represent patient's organizations; and three members nominated by EC based on EMA's recommendations. The assessment and review of orphan applications are with COMP but the final opinion and recommendation for approval is the responsibility of the CHMP.

Pediatric Committee

This is the committee in EMA responsible for evaluation and review of pediatric investigation plans. In cases where the applicant can justify no need for pediatric use for the product, a waiver or partial waiver may be granted by Pediatric Committee (PDCO). The applications may also be granted a deferral. PDCO members serve for 3 years and their appointments can be renewed. The committee is composed of the following members: five members of CHMP and their alternates appointed by CHMP; one member and alternate by each EU member state; three members and alternates representing health care professionals; three members and alternates representing patient organizations.

Coordination Group for Mutual Recognition and Decentralized Procedures

If there is disagreement between member states during the assessment of the submitted data for an MAA, on the grounds of a potential serious risk to public health, the CMD considers the matter and strives to reach an agreement within 60 days. If this is not possible, the member state responsible for the product brings the case to the attention of the CHMP for arbitration. The CMD is composed of one representative per member state (plus Norway, Iceland, and Liechtenstein), appointed for a renewable period of 3 years. Member states may also appoint an alternate member. Observers from the EC and EU accession countries also participate in meetings.

The Scientific Advice Working Party

The Scientific Advice Working Party (SAWP) is a standing working party that provides scientific advice, protocol assistance, and qualification of biomarkers. SAWP is established by the CHMP. The views of SAWP are provided to the CHMP and the COMP. The SAWP is a group comprised of members with different expertise who meet 11 times a year at EMA. The SAWP members include the following:

- a chairperson
- 28 members including three members of the COMP
- one member of the Committee for Advanced Therapies
- one member of the PDCO

In nomination of SAWP members, a fair representation of member states is ensured and also a representation of the following expertise on the working party must be ensured: pharmacokinetics; nonclinical safety;

methodology and statistics; and therapeutic fields for which there are frequent requests (e.g., cardiology, oncology, diabetes, neurodegenerative disorders, and infectious diseases).

ADAPTIVE PATHWAYS

In March 2014, the EMA announced a pilot program to launch "adaptive pathways," which used to be called "adaptive licensing" in the recent past [4]. In this pilot program, the EMA selected six drugs to move forward in its adaptive pathways program. The adaptive pathways seek to accelerate patients' access to medicines that are for treatment of serious unmet medical needs. Adaptive pathways allow medicines to obtain approval for use in limited indications or a limited (well-defined) patient subgroup prior to using the medicine in a broader indication or broader patient population.

Adaptive pathways are a logical product from years of experiencing risk-based treatment choices. In conventional drug approval process, every drug is unapproved and cannot be used in patients until it obtains an MAA. An exception was compassionate use where a physician could request a pharmaceutical company to provide their investigational drug (with no approval) for a particular named patient. The use in a named patient had to be justified based on several criteria such as lack of alternative treatments, failure of other approved treatments, and life-threatening medical condition for the patient. The adaptive pathways allow the use of a partially assessed product for a subgroup of patients as long as the risk–benefit ratio is favorable to the patient based on the available data. For example, an investigational drug that is still in development can be authorized for use for a subgroup of patients, as long as there is adequate justification based on available data to support the risk–benefit ratio. In other words, when there is life-threatening medical condition for a subgroup of patients and there is no other alternative treatment option left, it is justified to use a drug under investigation as long as there is minimal data available to support its potential benefit compared to the risk.

Adaptive pathways could become an effective approach to provide patients access to lifesaving treatments as soon as they are available even before full approval. This is a systematic approach to make medicines available to a group of patients, even at the early stage of their development, especially where there is an unmet medical need.

The adaptive pathways could be stimulating for pharmaceutical companies, which have been under pressure financially in the past decades, to stay profitable from the early stages of drug development.

AN OVERVIEW OF REGULATORY PROCESS IN OTHER FAST-GROWING MARKETS

The FDA and the EMA foundations are older than majority of the agencies in other countries. In many other regions, such as countries in South America, Africa, and the Middle East, there is no formal and full evaluation of the drug applications. The countries in the aforementioned regions rely mostly on NDA and MAA approvals by the FDA and the EMA. The evaluation of new drug approvals in these countries is normally limited to the need for a particular medicine compared to other available products in the country. The assessment of safety and efficacy of the product is mainly based on the NDA and MAA approval already obtained in the US and the EU [5,6].

In certain countries such as Japan, China, India, Canada, and Australia there is more formal assessment of drug applications and specific processes are involved. This section briefly highlights important differences and processes in Japan and China.

Japan

Traditionally, Japan has had a long lag time for approval of drugs that are approved elsewhere in the world [7]. In 2004, the statistics showed that while the US and majority of the EU countries take on average 500–620 days to approve a drug first approved elsewhere in the world, a similar approval would take about 1400 days in Japan. In recent years, Japanese authorities have revised processes to reduce this lag, but the gap still somewhat persists.

The Pharmaceuticals and Medical Devices Agency (PMDA) is the Japanese authority dealing with the review and operational aspects of regulatory drug development. The PMDA reports to the Ministry of Health, Labour and Welfare (MHLW). The MHLW is the final approver for a drug application and responsible for publishing guidelines and advisory committee. But assessments, evaluations, operational aspects (e.g., assessment of good clinical practice and good manufacturing practices (GMP) compliance), and interactions with the applicants are primarily with the PMDA. The PMDA is very highly concerned with safety. The PMDA normally requires bridging data in Japanese subjects regardless of the amount of data available in studies that are with Western or European population. The bridging data most often are in the form of a Phase 1 or Phase 2 study that would include Japanese subjects or patients in order to determine the safety and efficacy of the drug in Japanese population. The bridging data are required to enable extrapolation of the results from the original application (in the NDA or MAA), which is normally in non-Japanese, to Japanese subjects. The main aim of the bridging study is to ensure differences in Japanese subjects (compared to the non-Japanese) are identified and reflected in the product label, where applicable. Most often the design of the bridging study is similar to the study in the original approval.

By the nature of the process described above, the majority of pharmaceutical companies prefer to submit the applications to the PMDA after approval of a drug is obtained from the FDA or the EMA (or both). Therefore, bridging studies in Japan typically start long after FDA/EMA approval and typically take 1–2 years longer to complete. After testing the product in Japanese subjects, the drug application can be submitted to the PMDA for evaluation, which would take an additional 1–2 years with potential requests for further data or studies. This process creates a long lag for approval of new drugs in Japan typically 3–5 years after an FDA/EMA approval. Japanese authorities are conscious of this lag and in the past few years have attempted to come up with measures to address this issue.

More recently, Japan has taken the lead to initiate several processes to encourage and reduce drug lag time described above. Among these, Japan is encouraging pharmaceutical companies to initiate global clinical trials that include Japan as a region simultaneous or soon after completion of Phase 1 in non-Japanese subjects. Another initiative that is promoted by Japan is the collaboration between East Asian countries, mainly Japan, Korea, and China. Japan along with Korea and China have led a series of studies to show similarity in the ethnicity of East Asian populations as far as concerned with metabolism and efficacy of medicinal products. This would allow any clinical trial data collected from subjects with any East Asian ethnicity to be extrapolated to all East Asian countries including Japan. This in turn could facilitate and encourage pharmaceutical companies to include East Asian subjects and potentially result in removal of the requirement for doing bridging trials that are currently done exclusively in Japanese subjects. Overall, these initiatives could pave the way for pharmaceutical companies to include East Asian countries in their clinical trials early in the development of a drug and gather data on East Asian subjects that are applicable/acceptable to all East Asian countries including Japan, Korea, and China. Although these initiatives are steps in the right direction, they are in the early stages of progress and maybe a long time before any harmonized regulatory approach for East Asia is formalized. In practice, pharmaceutical companies still remain reluctant to include East Asia in the early clinical trials in parallel with trials in Western countries.

China

The drug approval process is modeled to some extent based on the US FDA. The China Food and Drug Administration (CFDA), formerly known as the State Food and Drug Administration (SFDA), evaluates medicinal products, that are approved and marketed in other countries, as new drugs in China [8]. Whether a drug is approved elsewhere in the world or not, the CFDA requires clinical data from trials performed in China to support an application. To do clinical trials, a CTA is required (similar to IND in the USA).

For drugs that are approved elsewhere in the world, a category III registration process must be followed, which results in an Import Drug License (IDL). The IDL application requires clinical trials in Chinese subjects. A pharmacokinetic study and a clinical study are usually required. Normally, the clinical study should include 100 pairs of subjects (100 on test treatment).

For drugs that are not approved in any other country yet, applicant is required to do a full clinical development program in China to obtain market approval. This is a category I registration process and by definition, the drug must not have been approved in any other country at the time of NDA in China.

Category III registration seems, at first glance, to be a faster path for drug registration. However, in practice, the drug developers experience 4–6 years lag between the product approvals in the US or the EU and the approval in China. This lag is due to major differences in approval time lines especially for CTA reviews by CFDA. In China, CTA review by CFDA takes on average a year and then an average of 1.5 year for conduct of clinical trials plus 1–1.5 years for IDL review to get a category III drug approval. In addition to the longer time lines, there are further challenges for using category III registration; the CFDA requires that the clinical trials must be conducted without any variation in protocol or any amendments. This may be delaying the recruitment of subjects (due to local differences in clinical practice) or compromise the outcome of the study.

In contrast to category III registration, a pharmaceutical company applying for a category I application could start global clinical trials (that include China). The company could submit an NDA in China (along with other parts of the world) when the clinical trials are completed globally. This would potentially allow drug registration and approval by CFDA in parallel (or with little lag) after approval by the FDA or the EMA. Therefore, pharmaceutical companies must carefully choose strategies for drug approval through CFDA. In majority of cases, a category I registration is faster than category III.

LIST OF ACRONYMS AND ABBREVIATIONS

ADME Absorption, distribution, metabolism, and excretion
BLA Biologics license application
cGMP Current good manufacturing practices
CBER Center for Biologics Evaluation and Research
CDER Center for Drug Evaluation and Research

CFR Code of Federal Regulations
CTA Clinical Trial Application
FDA Food and Drug Administration
IND Investigational New Drug
NDA New drug application
PAIs Preapproval inspections
PD Pharmacodynamic
PK Pharmacokinetic
US United States

References

[1] Cost to develop and win marketing approval for a new drug is $2.6 billion. Tufts Cost Study. http://csdd.tufts.edu/news/complete_story/pr_tufts_csdd_2014_cost_study=; November 18, 2014 [accessed on 01.06.15].

[2] European Medicines Agency website. http://www.ema.europa.eu/ema/index.jsp?curl=pages/about_us/general/general_content_000109.jsp [accessed on 01.06.15].

[3] European Medicines Agency website. http://www.ema.europa.eu/ema/index.jsp?curl=pages/about_us/general/general_content_000217.jsp&mid [accessed on 01.06.15].

[4] Adaptive Pathways. http://www.pmlive.com/pharma_intelligence/gathering_pace_adaptive_pathways_631869; February 3, 2015 [accessed on 03.06.15].

[5] Downing NS, Aminawung JA, Shah ND, Braunstein JB, Krumholz HM, Ross JS. Regulatory review of novel therapeutics – comparison of three regulatory agencies. N Engl J Med 2012;366:2284–93.

[6] Howie LJ, Hirsch BR, Abernethy AP. A comparison of FDA and EMA drug approval: implications for drug development and cost of care. Oncology (Williston Park) 2013;27(12):1195. 1198–1200.

[7] Tsukamoto E, Tripathi S. Japan's drug lag and national agenda. Regulatory Focus; 2011. p. 35–41.

[8] Website. http://eng.cfda.gov.cn/WS03/CL0755/ [accessed on 03.06.15].

CHAPTER

6

Translational Medicine Case Studies and Reports

Alexandre Passioukov[1], Pierre Ferré[2], Laurent Audoly[2]

[1]Head of Translational Medicine, Pierre Fabre Pharmaceuticals, Toulouse, France; [2]Pierre Fabre Pharmaceuticals, Toulouse, France

INTRODUCTION

Identification and preclinical validation of the drug target followed by clear clinical proof of concept is central to translational medicine methodology. Perhaps more than any other therapeutic area so far, this has been achieved in the field of oncology, where the "bench-to-bedside" loop has led to significant improvements in the treatment of numerous cancer types, often with dramatic results for patients refractory to existing therapies. In this chapter, we present three progressively multilayered examples of how the interplay between fundamental scientific discoveries and clinical experience has driven the development of innovative cancer treatments over the past two decades.

Historically, hematologic malignancies have been the privileged indications for targeted therapeutic approaches. Typically, the target is highly expressed on the surface of neoplastic cells present in large numbers in the blood and secondary lymphoid organs and therefore is readily accessible to systemically administered agents. As well, malignant cells can be easily sampled from the patient, both for pharmacodynamic assessment of the response and to provide mechanistic insights. The first case study describes the development of rituximab (Rituxan®), a chimeric human–mouse monoclonal antibody (mAb) against the B cell lineage marker CD20, present on many lymphomas. As CD20 is inherent to the disease process itself there is no requirement for a dedicated test to select potential responders to anti-CD20 therapy. Prior to rituximab's approval by the US Food and Drug Administration (FDA), treatment of advanced or relapsed B cell lymphomas relied on aggressive chemotherapy agents or radiotherapy, often with significant side effects. Due to their high specificity for tumor cell targets, mAbs had long been considered as ideal anticancer agents, although development of the first murine anti-CD20 mAbs was hampered by immunogenicity problems. Rituximab was the first mAb to be approved specifically for cancer therapy [1] and in recent years, research has focused on identifying the molecular mechanisms underlying rituximab's therapeutic action with the aim of developing novel agents with enhanced therapeutic profiles.

Compared with hematologic malignancies, targeted treatment of solid tumors presents a far more complex situation. Cellular heterogeneity within a given tumor type presents a significant challenge in terms of designing effective treatment strategies, and even if a suitable therapeutic target is identified, drug delivery to malignant tissue is often hindered by architectural and physiological features of the tumor [2]. Our second case highlights one of the major success stories of translational medicine in oncology in the past decade. Here, the discovery that a small minority of patients with non-small cell lung cancer (NSCLC) expressed chromosomal rearrangements of anaplastic lymphoma kinase (ALK) led to rapid clinical development of the tyrosine kinase inhibitor (TKI) crizotinib (Xalkori®) along with its companion diagnostic. The ALK rearrangement is an "ideal" solid tumor marker for targeted treatment: as an oncogenic driver it is expressed by the vast majority of tumor cells and is technically easy to identify using standard diagnostic techniques. Yet, despite exceptional response rates obtained in ALK-positive NSCLC, drug resistance to crizotinib eventually develops in many patients. Therefore, recent translational research efforts have concentrated on developing strategies to overcome resistance mechanisms. Beyond NSCLC, increasing translational knowledge gained from the systematic molecular analysis of diverse tumors has aided identification of subsets of patients with other histologies who are also likely to benefit from ALK-TKI treatment.

The third case concerns cancer immunotherapy, which is emerging as a powerful and innovative approach to the treatment of advanced solid tumors. Following several decades of basic immunology research, scientific breakthroughs in the mechanisms responsible for tumor immunity enabled the clinical development of antibodies directed at "immune checkpoint receptors." Several agents, including antagonist mAbs to cytotoxic T lymphocyte-associated antigen 4 (CTLA-4) (e.g., ipilimumab, Yervoy®) and programmed death 1 (PD-1) (e.g., pembrolizumab, Keytruda®; nivolumab, Opdivo®), have already been approved for the treatment of advanced or unresectable melanoma, with other indications currently under regulatory review. Such treatments involve a completely unique approach to attacking tumor cells, centered on interfering with the cross talk between the patient's immune system and the immunosuppressive tumor microenvironment. Although the biological complexity of such an approach is high, in turn we are rewarded with durable and potentially curative responses. Nevertheless these are currently limited to sensitive indications and subsets of patients, and translational approaches are now key to the development of rational immunotherapy combinations to treat a broader range of malignancies.

CASE STUDY 1: MABS TARGETING CD20-POSITIVE HEMATOLOGIC MALIGNANCIES

CD20 was discovered as a B cell marker in 1980 [3] and is highly expressed in the plasma membrane of normal B cells, from early pre-B cell development until the activated B cell stage. It is also expressed by most B cell lymphomas and some leukemias [4], including chronic lymphocytic leukemia (CLL). Indeed, because of its broad proliferation in malignancy, CD20 presented a highly attractive target for antibody-mediated therapy. The concern, however, was that targeting such a marker would not only lead to depletion of malignant cells but also normal B cells and plasma cell progenitors, potentially disrupting antibody production. During initial studies in cynomolgus monkeys, repeated rituximab treatment caused marked B cell depletion from peripheral blood (and lymph nodes, spleen, and bone marrow), with no evidence of toxicity [5]. These preclinical findings were subsequently translated into the clinic: in a Phase 1 trial in which 15 patients with relapsed low-grade B cell lymphoma received a single intravenous dose of rituximab (ranging from 10 to $500 \, mg/m^2$), CD20-positive B cells were rapidly depleted in the peripheral blood at 24–72 h postinfusion and remained depleted for

continued

CASE STUDY 1: MABS TARGETING CD20-POSITIVE HEMATOLOGIC MALIGNANCIES *(cont'd)*

several months in most cases [6]. No dose-limiting toxicities or immune responses were observed, and clinically significant tumor regression occurred in 6 of 15 patients. In a Phase 2 trial in patients with relapsed low-grade lymphoma or follicular lymphoma (FL), clinical responses were observed in 46% of patients treated with four weekly rituximab infusions at a dose of 375 mg/m^2 [7]. Treatment was well tolerated (most adverse events were mild infusion-related reactions after the initial infusion) and caused rapid depletion of peripheral B cells, without clinically significant changes in serum Ig levels or an increase in the frequency or severity of infections. A similar (48%) response rate was observed in a Phase 2 confirmatory study [8] and this ultimately led to FDA approval of rituximab for relapsed or refractory low-grade B cell lymphoma or CD20-positive FL, notably before any formal comparative Phase 3 trial had been performed. Since first approval in 1997, rituximab has become the cornerstone of management of many B cell malignancies. With its favorable toxicity profile, rituximab can be safely used in combination with conventional chemotherapy in low-grade lymphomas and aggressive diffuse large B cell lymphoma [9] and as single-agent maintenance therapy in FL [10].

Although targeting CD20 for the treatment of CD20-positive malignancies is conceptually straightforward, the underlying biology is decidedly more complex. As discussed below, several mechanisms may explain rituximab's clinical activity but their relative contribution in the clinical setting is unknown. Moreover, despite rituximab's clinical success, a substantial proportion of patients with B cell lymphomas are unresponsive to first treatment (primary resistance) or eventually develop resistance to treatment over time (acquired drug resistance). Here, translational research can improve on our existing knowledge of the molecular pathways of clinical response and resistance, enabling development of more potent therapeutics around the same or new relevant molecular targets.

In terms of mode of action, a large body of preclinical and clinical evidence suggests that rituximab and other anti-CD20 mAbs can deplete B cells via diverse effector pathways, involving Fc-dependent processes, namely, antibody-dependent cellular cytotoxicity (ADCC) and complement-dependent cytotoxicity (CDC), and also programmed cell death (PCD) [11]. While a role for CDC and PCD is disputed, interactions between the Fc domain of the anti-CD20 mAb and IgG Fc-gamma receptors (FcγR) expressed on effector cells (e.g., monocytes, macrophages,

CASE STUDY 1: MABS TARGETING CD20-POSITIVE HEMATOLOGIC MALIGNANCIES *(cont'd)*

neutrophils, and natural killer (NK) cells) are seemingly important in the antitumor effects of anti-CD20, and were initially demonstrated pre-clinically in murine models [12]. More recently, genotyping of samples from patients in clinical studies indicates that some FcγR polymorphisms are associated with better clinical responses to rituximab therapy in FL [13,14] but not in other malignancies such as CLL [15], suggesting that the relative involvement of effector mechanisms differs according to B cell lymphoma subtype. The anti-CD20 mAb response may also involve T cell-mediated immunization, potentially explaining the durable clinical responses seen after a single course of rituximab [16]. Specifically, preclinical work showed that rituximab-treated lymphoma cell lines were able to promote cross-priming of CD8+ cytotoxic T cells against lymphoma antigens [17]. Later, in a proof-of-principle study, an increase in idiotype-specific T cells occurred after rituximab treatment in 4 of 5 patients with FL, indeed supporting a "vaccinal effect" of rituximab [18]. Future translational research will hopefully offer further insight into the importance of this mechanism.

Numerous mechanistic pathways for rituximab resistance have been proposed but their contribution to the clinical picture remains unclear [11,19]. Expression of membrane complement regulatory proteins (e.g., CD55 and CD59) by tumor cells might confer protection against CDC after repeated exposure to rituximab [20]. Alternatively, loss of surface CD20 expression due to transcriptional downregulation [21] and internalization of CD20–mAb complexes by targeted B cells [22,23] may contribute to rituximab resistance. More recent evidence suggests that trogocytosis plays a role; in this process, bound CD20–mAb complexes are "shaved" from the target cell surface by phagocytic Fc-bearing cells (rather than being internalized), allowing malignant cells to escape. This is thought to occur when effector mechanisms necessary for mAb activity (in this case phagocytosis) become saturated under conditions of high tumor burden. Trogocytosis has been proposed as a potential mechanism to partly explain the poor clinical responses to rituximab observed in many patients with CLL, a disease characterized by high numbers of peripheral malignant cells [24]. More specifically, clinical investigation showed that while clearance of peripheral B cells was rapid following a standard infusion of rituximab in patients with CLL, a B cell population with significantly reduced levels of CD20 reappeared in the bloodstream soon afterward [25]. Loss of CD20 from B cells was later confirmed in

continued

CASE STUDY 1: MABS TARGETING CD20-POSITIVE HEMATOLOGIC MALIGNANCIES *(cont'd)*

in vitro and in vivo models of the shaving process [26,27] Importantly, a small pilot study suggested that rapid clearance of CD20+ cells could be achieved even at very low doses of rituximab [28] and the authors proposed that a low-dose strategy to reduce trogocytosis may be more effective than standard dosing for treatment of CLL [24]. Conceivably, anti-CD20 mAbs specifically engineered to avoid trogocytosis may provide more effective CLL treatments.

Improved understanding of the mechanism of action of rituximab has led to the development of novel agents aimed at further enhancing efficacy. Anti-CD20 mAbs are broadly classified as Type I or Type II, based on their differing mechanisms of action in vitro. Type I mAbs (including rituximab) redistribute CD20 into lipid rafts and induce CDC, whereas Type II mAbs are characterized by their ability to evoke PCD [29]. Ofatumumab, a fully human Type I mAb, binds to a novel epitope on CD20 and has a slower dissociation rate and markedly enhanced CDC compared with rituximab [30]. On the other hand, obinutuzumab (GA101), a humanized Type II mAb, was glycoengineered to produce enhanced ADCC (via modification of the Fc region) and direct induction of PCD [31]. Ofatumumab and obinutuzumab have both been approved for use in CLL, although whether these and other next-generation anti-CD20 mAbs are more efficacious than rituximab remains to be established. In one study, obinutuzumab was shown to be superior to rituximab when each mAb was given in combination with chlorambucil in CLL patients, although the higher dose of obinutuzumab used precludes any conclusion about increased clinical benefit over rituximab in this case [32]. Indeed, the different dosing schedules used for various mAbs in clinical trials make it difficult to ascribe any differences in clinical activity to mechanistic differences between agents [33].

In summary, clinical development of rituximab as the first targeted anti-CD20 mAb marked the beginning of a new era in cancer therapy, and nearly two decades after its introduction, rituximab is still the first-line treatment for many B cell malignancies. Recent translational efforts have focused on dissecting out the mechanism of action of anti-CD20 mAbs and several novel agents have been approved. Nevertheless, fundamental questions remain about its precise mode of action, optimal treatment regimens, and why some patients are resistant to therapy. From a translational perspective there is still work to be done to further improve on current clinical outcomes and enhance our mechanistic understanding of B cell disorders and similar malignancies.

CASE STUDY 2: SMALL MOLECULES TARGETING ALK IN LUNG CANCER AND THE COMPANION DIAGNOSTIC APPROACH

Cytotoxic chemotherapy is the frontline treatment for many patients with advanced NSCLC but is typically associated with only modest response rates and small improvements in overall survival (OS) [34]. Research into the molecular drivers of NSCLC provided the foundation for targeted therapy in genetically defined patient populations, beginning with the discovery that a small subset of patients with activating mutations in the epidermal growth factor receptor (EGFR) gene were highly responsive to treatment with the TKIs, gefitinib and erlotinib [35–37]. ALK was subsequently identified as a potential therapeutic target in NSCLC in 2007. Researchers found that a small chromosomal inversion causes fusion of the ALK and echinoderm microtubule-associated protein-like 4 (EML4) genes [38]. The resulting *EML4–ALK* gene rearrangement causes expression of a constitutively activated tyrosine kinase with potent in vitro and in vivo oncogenic activity [38,39] and is present in around 2–5% of NSCLC tumors [40], particularly in patients with light smoking history, younger age, and adenocarcinoma histology [41–43]. Since the discovery of the EML4–ALK rearrangement, other less frequent fusion partners for ALK have been described, including KIF5B [44] and TFG [45]. Indeed, due to the collective efforts of many research groups the descriptive landscape of molecular aberrations in lung cancer and other malignancies has been greatly improved [46].

Identification of ALK as a putative NSCLC target serendipitously coincided with the development of crizotinib (formerly PF-2341066), an orally bioavailable small molecule TKI, which was being investigated for other ALK- or c-MET-driven cancers [47,48]. Crizotinib was initially evaluated in a Phase 1 dose-escalation study in patients with advanced solid tumors, beginning in 2006. Notably, the drug showed clinical activity in two patients with NSCLC who were positive for the ALK rearrangement, detected as a result of the protocol-specified screening of patient tumor samples for the presence of ALK or MET activation [49]. Consequently, the study protocol was amended in 2008 to include an additional expanded cohort of patients with ALK-positive stage III/IV NSCLC. Of approximately 1500 patients screened for ALK-positive tumors, 82 patients were detected, selected, and treated with crizotinib (250 mg twice a day in 28-day cycles). The overall response rate in those patients was high (57%), and a further 33% had stable disease [49]. The drug was well tolerated with mild gastrointestinal side effects as the most common

continued

CASE STUDY 2: SMALL MOLECULES TARGETING ALK IN LUNG CANCER AND THE COMPANION DIAGNOSTIC APPROACH (cont'd)

adverse event. In an updated analysis of this Phase 1 study (involving 149 patients), 61% had an objective response and median progression-free survival (PFS) was 9.7 months [50]. Subsequently, a Phase 2 study employing the same dose regimen of crizotinib showed similarly impressive response rates (59.8%) and a median PFS of 8.1 months [51]. On the basis of the Phase 1/2 results, crizotinib was granted accelerated approval by the FDA for the treatment of locally advanced or metastatic NSCLC that is ALK-positive by an FDA-approved test. Thus, as with rituximab, crizotinib was approved in the absence of Phase 3 data, although in this case the time frame between discovery of the target (2007) and drug approval (2011) was considerably shorter.

Crizotinib can be considered as a "model" targeted agent, particularly since the companion diagnostic to identify ALK-positive patients was codeveloped alongside the drug. The Vysis ALK break-apart fluorescence in situ hybridization (FISH) probe kit, used in the first crizotinib trial, was concurrently approved with crizotinib by the FDA. Currently, it is the only validated test to identify potentially responsive patients, although several other diagnostic assays are available or in development [52]. The ALK FISH assay is highly sensitive and specific, and can confirm the presence of all ALK rearrangements, irrespective of the fusion partner or variant. Nevertheless, the test is expensive, rendering it less suitable for high-throughput analysis of large numbers of patient samples, especially for detection of rare gene mutations and rearrangements present in only a small percentage of patients. Clearly, the development and application of alternative companion diagnostics using multiplex technologies to reduce the amount of patient material required will likely improve detection of ALK and other clinically relevant tumor-specific aberrations.

Besides NSCLC, ALK rearrangements have been detected in other tumor types, and some of these appear to be responsive to crizotinib therapy, particularly inflammatory myofibroblastic tumors [53] and anaplastic large cell lymphomas [54]. This highlights one of the true strengths of translational medicine research and presents a unique paradigm whereby identification of the molecular target (in this case ALK rearrangement) in a small number of patients and development of the first targeted agent leads to prevalence analysis of the target and subsequent use of this agent in other indications, irrespective of histology. For example, with respect to crizotinib, which is currently registered only for the treatment

CASE STUDY 2: SMALL MOLECULES TARGETING ALK IN LUNG CANCER AND THE COMPANION DIAGNOSTIC APPROACH (cont'd)

of patients with ALK-positive lung cancer, a French nationwide initiative was launched in 2013 to allow controlled access to the drug for patients with ALK-, MET-, or ROS1-positive tumors [55]. The program is now being expanded to facilitate access to other targeted agents that are potentially helpful for the treatment of molecularly selected subsets of patients (beyond the original registration label), for whom there are no registered treatments available.

As with many molecularly targeted treatments, crizotinib's therapeutic efficacy is limited by acquired drug resistance: most ALK-positive patients with NSCLC eventually experience disease progression within a year of starting treatment. Several mechanisms have been described, most frequently secondary ALK mutations, as well as ALK gene amplification, activation of alternative signaling pathways (e.g., EGFR and KIT), and other, as yet, unidentified mechanisms [40]. As with rituximab, translational research has played a crucial role in providing mechanistic insights into crizotinib resistance and even before the drug was registered, the first published reports of resistance mechanisms had already emerged. Using deep sequencing techniques on pleural fluid samples, Choi et al. [56] discovered two point mutations in EML4–ALK, which conferred resistance to crizotinib when introduced into an EML4–ALK-positive BA/F3 mouse cell line. Protein structure modeling revealed that one of the mutations acted as a "gatekeeper" resistance mutation, causing allosteric hindrance of crizotinib binding to the kinase domain-binding site. Translational efforts such as these, involving comprehensive and specific molecular analysis of patient samples and correlation with clinical outcomes, very rapidly led to the development of numerous second-generation ALK inhibitors. For example, ceritinib (Zykadia®) is a highly potent and selective ALK inhibitor that inhibits several (but not all) crizotinib-resistant mutations in preclinical models [57]. This agent gained FDA approval in 2014 as a second-line treatment for patients with ALK-positive NSCLC with failure or intolerance to crizotinib. Impressively, approval was based solely on Phase 1 results that showed similar response rates in crizotinib-naïve and crizotinib-treated patients, including those with known resistance mutations in ALK [58]. Alectinib is another highly potent selective ALK inhibitor with activity against crizotinib-resistant ALK mutations [59], and Phase 3 trials are currently in progress. As crizotinib apparently shows poor central nervous system

continued

CASE STUDY 2: SMALL MOLECULES TARGETING ALK IN LUNG CANCER AND THE COMPANION DIAGNOSTIC APPROACH (cont'd)

(CNS) activity, drugs with improved blood–brain barrier penetration might improve outcomes in patients with ALK-positive NSCLC and brain metastases. To this end, PF-06463922, a combined ALK and ROS1 inhibitor, has been specifically designed for enhanced CNS penetration and shows excellent antitumor activity in murine models of EML4–ALK-driven brain tumors [60]. This drug is currently in Phase 1/2 clinical trials in ALK-positive and ROS1-positive NSCLC. Other potential approaches for treating crizotinib-resistant disease include combining ALK inhibitors with drugs targeting EGFR (e.g., erlotinib) or KIT, or pharmacological blockade of heat shock protein 90, which regulates the stability of the EML4–ALK fusion protein [61,62].

Thus, the ALK–crizotinib story highlights the importance of early identification of the target and translation of basic research findings into the clinic, which in this case contributed to the rapid market approval of a new targeted treatment along with its companion diagnostic. It also shows how knowledge of mechanisms of acquired resistance to the first-in-class agent can be rapidly employed for the development of therapeutically superior agents to bypass resistance.

CASE STUDY 3: IMMUNO-ONCOLOGY TRANSLATIONAL STORY—FROM IPILIMUMAB IN MELANOMA TO CTLA-4/PD-1/PD-L1 COMBINATIONS AND BEYOND

The field of immuno-oncology was transformed by the landmark discovery that tumor cells cause local immune suppression through dysregulation of immune checkpoint receptors expressed on the surface of T cells, effectively shielding themselves from host immunity. Since the discovery of the first checkpoint receptor CTLA-4, large families of T cell costimulatory and coinhibitory receptors have been identified, which under normal physiological conditions are critical for maintenance of immune tolerance and preventing autoimmunity. However, immune checkpoint receptors and their ligands can also be aberrantly expressed

CASE STUDY 3: IMMUNO-ONCOLOGY TRANSLATIONAL STORY—FROM IPILIMUMAB IN MELANOMA TO CTLA-4/PD-1/PD-L1 COMBINATIONS AND BEYOND *(cont'd)*

by tumor and tumor-associated cells as a mechanism to confer immune resistance. Collectively, this knowledge is the product of many decades of basic and translational research into cancer immunology, and the long-held dream of applying it to cancer therapeutics was finally realized when the first agents, mAbs directed at CTLA-4 and subsequently PD-1, were approved for the treatment of advanced melanoma. Indeed, with impressive responses observed in many patients with metastatic disease, we have witnessed a remarkable paradigm shift from palliative to potentially curative treatment. Consequently, this area is the most exciting and rapidly advancing field in cancer research today.

The coinhibitory receptor CTLA-4 is expressed on T cells where it serves to downregulate the amplitude of early T cell activation. CTLA-4 counteracts the activity of the costimulatory receptor CD28 following antigen recognition [63,64] and plays an essential role in maintaining immunological homeostasis. Seminal preclinical work published in 1995 showed that in vivo administration of antibodies to CTLA-4 caused rejection of some preestablished transplantable murine tumors [65], raising the possibility that mechanisms of T cell dysfunction in the tumor microenvironment could be therapeutically manipulated to reengage the immune system and restore tumor immunity. This led to clinical development of mAbs directed against CTLA-4. Ipilimumab, a fully human IgG1 anti-CTLA-4 mAb, was evaluated in a clinical trial program involving >2000 patients with a variety of solid tumors. This mAb gained US FDA approval in 2011 for the treatment of unresectable or metastatic melanoma, based on data from two Phase 3 randomized trials, which showed improved OS versus control arms [66]. Even more noteworthy was the improvement in long-term survival, with nearly 20% of patients surviving >2 years despite the relatively short treatment course. Ipilimumab has also shown positive results in other tumor types, particularly NSCLC, small cell lung cancer, bladder cancer, and metastatic prostate cancer.

PD-1, another key coinhibitory immune checkpoint receptor, regulates effector T cell activity during inflammation and dampens autoimmunity. In addition to T cells, PD-1 is expressed by other activated immune cell types, including B cells and NK cells [67]. In many malignancies, tumor cells and immune cells commonly express PD-1 ligands, particularly programmed death ligand-1 (PD-L1), and tumor-infiltrating lymphocytes

continued

CASE STUDY 3: IMMUNO-ONCOLOGY TRANSLATIONAL STORY—FROM IPILIMUMAB IN MELANOMA TO CTLA-4/PD-1/PD-L1 COMBINATIONS AND BEYOND (cont'd)

(e.g., CD4+ and CD8+ T cells) express PD-1 and PD-L1 [67–69]. Therapeutic strategies to block the PD-1/PD-L1 pathway include anti-PD-1 antibodies designed to inhibit PD-1 from engaging with its ligands, or antibodies directed against PD-L1, which prevent the ligand from engaging with PD-1. Pembrolizumab and another agent nivolumab (both anti-PD-1 mAbs) were both approved in 2014 for the same indication as ipilimumab under the FDA's accelerated approval program [70]. Clinical development of both agents was extremely rapid, with pembrolizumab taking only 3 years from the first clinical trials to approval. Recently, several PD-L1 inhibitors have reached late-stage clinical development and appear to have impressive antitumor effects in a number of malignancies.

While durable responses to immune checkpoint inhibition can be achieved, these effects are currently limited to patient subsets. As a direct result, translational research investigations were established at a very early stage to increase the number of patients that could potentially benefit from these agents. More specifically, two main translational approaches have been employed in parallel to guide the design and development of rational combination strategies. The first approach has relied on data from preclinical models—the so-called "bench-to-bedside" concept. Early knockout mice studies indicated that T cell inhibition by CTLA-4 and PD-1 occurred largely via nonoverlapping pathways, suggesting that simultaneous CTLA-4 and PD-1/PD-L1 blockade might be necessary to treat particularly aggressive tumors [71]. Indeed, triple blockade of CTLA-4, PD-1, and PD-L1 in mice expressing the B16–BL6 melanoma clone caused a higher tumor rejection rate compared to combination blockade of CTLA-4/PD-1 or blockade of any one [72]. In the clinic, combination CTLA-4/PD-1 blockade with ipilimumab/nivolumab led to rapid and deep tumor regression in 9 of 17 patients (53%) with advanced melanoma in a Phase 1 study [73]. Long-term follow-up showed impressively high survival rates (94% and 88% at 1 and 2 years, respectively), suggesting that a significant proportion of those patients could be permanently cured [74]. In light of these encouraging results in melanoma, clinical trials are now testing this strategy in other tumor types.

In addition to the CTLA-4 and PD-1/PD-L1 pathways, agents modulating other novel immune checkpoint molecules are now under active clinical development, including numerous checkpoint inhibitors as well

CASE STUDY 3: IMMUNO-ONCOLOGY TRANSLATIONAL STORY—FROM IPILIMUMAB IN MELANOMA TO CTLA-4/PD-1/PD-L1 COMBINATIONS AND BEYOND *(cont'd)*

as several agonistic antibodies directed toward the T cell costimulatory receptors 4-1BB, OX-40, GITR, and ICOS, among others [71]. The therapeutic potential of combining CTLA-4 blockade with 4-1BB activation to restore tumor immunity has already been demonstrated preclinically [75–77]. Several 4-1BB agonist mAbs entered clinical trials, although one agent was associated with an increased risk of life-threatening liver toxicity in some patients [78]. Preclinical evidence, however, suggests that CTLA-4 blockade can ameliorate the hepatotoxic effects of 4-1BB agonist treatment, and that 4-1BB activation can block the autoimmunity arising from blockade of CTLA-4 [77]. Indeed, an important consideration with any treatment approach involving immune checkpoint modulation is the risk of immune-related toxicities. In patients treated with combined nivolumab/ipilimumab, the spectrum of adverse events reported was similar to previous experience with nivolumab or ipilimumab monotherapy but grade III/IV treatment-related events were reported in over half of the patients [73].

The second main translational approach in place focuses on deciphering the mechanisms of resistance to anti-CTLA-4 or anti-PD-1/PD-L1 treatments currently in clinical evaluation, achieved via detailed analysis of patient-derived samples ("bedside-to-bench" translation). Most clinical protocols are (or at least should be) designed to strengthen the efforts of sequential sampling over the whole treatment duration, ideally including tumor biopsies from responders and nonresponders to single immune checkpoint blockade. The long-term goal here is identification of the mechanisms of immune escape or resistance, ultimately enabling selection of the optimal drug combination partners for future preclinical and clinical investigations.

Another intense area of translational research is the identification of predictive biomarkers to facilitate the selection of patients most likely to benefit from checkpoint modulation (and to exclude those with an increased risk of immune-related adverse events), although clinically applicable biomarkers have yet to be identified. PD-L1 expression via immunohistochemistry (IHC) has been widely investigated as a biomarker for response to PD-1/PD-L1 pathway blockade and studies across multiple histologies generally indicate higher response rates and OS in PD-L1-positive patients [68]. However, cross-study comparison

continued

CASE STUDY 3: IMMUNO-ONCOLOGY TRANSLATIONAL STORY—FROM IPILIMUMAB IN MELANOMA TO CTLA-4/PD-1/PD-L1 COMBINATIONS AND BEYOND *(cont'd)*

is limited by many factors, such as differences in the type of assay and detection antibodies used, PD-L1 positivity cutoffs, and tissue preparation and cell sampling techniques. In metastatic bladder cancer, PD-L1 expression on tumor-infiltrating cells may have a stronger predictive value to anti-PD-L1 treatment with MPDL3280A than tumor cell PD-L1 expression, indicating that enrichment of responders within this population based on this marker is achievable [79]. Interestingly, tumor PD-L1 status in advanced melanoma does not appear to predict therapeutic outcome in patients on combined ipilimumab/nivolumab treatment, as evidenced by similar tumor response rates in patients with PD-L1-positive and PD-L1-negative tumors [80].

With respect to anti-CTLA-4 therapy, several pharmacodynamic biomarkers appear to correlate with clinical response to ipilimumab, such as sustained ICOS expression on CD4+ effector T cells [81] or an increase in the absolute lymphocyte count [82]. In addition, high pretreatment expression of immune-related genes (e.g., FoxP3) and an increase in tumor-infiltrating lymphocytes may have predictive value for ipilimumab treatment [83]. This has led to the suggestion that immunoprofiling of the patient's genetic signature could be used to predict response to therapy [84]. One study showed that baseline diversity richness of the peripheral T cell receptor repertoire (assessed with a multiplex quantitative PCR technology) was associated with clinical benefit of ipilimumab [85]. Consequently, the kit is now being evaluated in larger patient cohorts to determine whether it can be used to select patients likely to respond to ipilimumab in metastatic melanoma and possibly other cancer immunotherapies.

To maximize clinical benefit across a broader range of patients, current research efforts are focusing on sequential or combination treatment approaches involving checkpoint inhibition, traditional therapies (radiotherapy, cytotoxic chemotherapy), vaccines, molecularly targeted agents, or investigational treatments [86]. Future strategies may utilize modalities to enhance antigen presentation or augment the number of tumor-specific T cells using vaccines or adoptive cell transfer [71]. Alternatively, agents directed against tumor myeloid cells (which form a barrier to T cell-driven immunotherapy) may be useful in combination with checkpoint modulation. The high biological complexity of interactions between the tumor and the patient's immune system highlights the need for "personalized treatment approaches" to cancer therapy, and likely different treatment combinations will be necessary even within the same patient segment. Deciphering immunosuppression patterns developed by individual tumors will be critical for the rationalized design of optimal treatment combinations, the development of which will require relevant preclinical models together with carefully designed clinical studies employing optimal dosing and treatment sequences. Conceivably, selection of patients most likely to benefit from single-agent versus combination therapy will rely on multiplex panels of biomarkers utilizing different technologies to identify *all* potential responders (while excluding the nonresponders). As a result, translational medicine is now entering a new phase involving highly complex challenges that will require unprecedented integration of translational and clinical research.

CONCLUSIONS

The examples presented above from the field of oncology demonstrate how translational medicine has proven to be a powerful tool for driving advances in biological science and the development of innovative therapeutic agents, with a major impact on clinical practice. The most straightforward situation in translational medicine involves identification and preclinical validation of the drug target, followed by clinical proof of concept in the clinic. On the other hand, through analysis of patient samples and knowledge generated from clinical trials, translational research can identify additional novel targets for the development of new active agents. Another crucial facet of translational medicine is the early identification of predictive biomarkers to enable the timely codevelopment of companion diagnostic assays.

The last two decades have witnessed a rapid growth in the number of translation studies in oncology, and consequently a vast amount of molecular data have been generated. Indeed, improved insight into cancer biology has led to the identification of many new molecular segments within formerly unique oncology indications, thus signaling the end of the "one histology—one drug" era. Accordingly, clinical trial design is now evolving to incorporate patient screening for molecular markers to predict response, independent of tumor histology. A future challenge will be the development of validated molecular screening platforms to bring this approach into routine clinical practice. With respect to immuno-oncology, robust identification of potential responders is likely to involve a combination of different markers, even for single-agent therapy, necessitating a shift from single to multiplex formats for guiding treatment decisions and monitoring treatment effects.

With the evolution of new technologies, invasive tumor biopsies may be progressively supplanted by peripheral surrogates from noninvasive liquid biopsies (e.g., circulating tumor DNA, tumor-derived materials from exosomes, etc.), and conceivably innovative imaging techniques will be among the next steps to revolutionize personalized medicine strategies for cancer treatment. The increasingly interdisciplinary nature of translational medicine highlights the need for global initiatives to further facilitate exchanges between academia, the pharmaceutical industry, and regulators. In addition, integrated databases of biomedical informatics, as well as more standardized methodology (e.g., biomarker assays) to allow cross-study comparisons are becoming a necessity. Clearly, clinical trials involving global molecular screening of a large number of patients will be particularly important for addressing smaller treatment indications, necessitating streamlined regulatory pathways for drug registration.

In summary, translational research effort can be viewed as a cyclic, iterative process, from laboratory bench to patient bedside, and the reverse. Improved understanding of tumor biology leads to the development of new therapeutic strategies, and tailoring of new agents toward responsive patients (often with codevelopment of diagnostic tools along the way) can ultimately bring ever-deeper mechanistic insights into the underlying disease biology. Thus, biomedical research today is virtually driven by the interplay between basic science and clinical reality and is moving with unprecedented speed, aided by "professionalized" translational medicine approaches. Indeed, while the field of oncology is a pioneer, other pathologies (e.g., cardiovascular and infectious diseases, immunological disorders) have already benefitted and will continue to benefit by following similar "translational pathways."

Glossary

ALK fusion oncogene Hybrid gene formed as a result of a structural rearrangement of the ALK gene that leads to fusion with a partner gene (e.g., EML4, NPM, KIF5B, or TFG); the resulting fusion gene is a driver of cancer cell proliferation.

Antibody-dependent cellular cytotoxicity Process in which Fc-gamma receptors (FcγR) on the surface of immune effector cells (e.g., natural killer cells, macrophages, monocytes, and eosinophils) bind the Fc region of a (therapeutic) antibody, itself specifically bound to a target cell. The Fc–FcγR interaction causes effector cell activation, resulting in the secretion of cytotoxic substances that mediate destruction of the target cell.

Biomarker Biologic characteristic (e.g., molecule, gene, etc.) that can be objectively measured as an indicator of normal biological or pathogenic processes, or pharmacologic responses to a therapeutic intervention.

CD20 B cell-specific differentiation antigen expressed on mature B cells (but not on early B cell progenitors or later mature plasma cells) and in many B cell non-Hodgkin's lymphomas.

Chimeric antibody Antibody made by fusing the antigen-binding region (variable domains of the heavy and light chains) from one species with the constant domain (effector region) from another species.

Companion diagnostic Test (such as an in vitro assay or measurement) to identify a predictive biomarker allowing selection of patients in whom the treatment is likely to provide benefit.

Complement-dependent cytotoxicity Process in which the classical complement activation pathway is triggered by binding and fixation of C1q protein to the Fc region of a (therapeutic) antibody; this leads to formation of the membrane attack complex at the surface of the target cell, ultimately causing cell lysis and death.

CTLA-4 (cytotoxic T lymphocyte-associated antigen 4) Coinhibitory T cell receptor that functions as a negative regulator of T cell activation, counteracting the activity of CD28. CTLA-4 binds the B7 family of costimulatory molecules and was the first immune checkpoint receptor to be discovered.

Fc-gamma receptor Receptor for the Fc portion of IgG expressed on many immune effector cells.

Fluorescence in situ hybridization Cytogenetic technique used to detect and localize the presence or absence of specific DNA sequences on chromosomes.

Humanized antibody Antibody made by grafting hypervariable regions of nonhuman antibodies into human antibodies, resulting in a molecule of approximately 95% human origin.

Immune checkpoints Receptors that regulate immune pathways (inhibitory or stimulatory) that play a crucial role in maintaining immune tolerance and modulating physiologic immune responses.

Immunohistochemistry Technique to identify discrete tissue components by the interaction of target antigens with specific antibodies tagged with a visible label.

Immunotherapy Therapeutic agent that can stimulate, enhance, or suppress the immune response in order to fight disease.

Monoclonal antibody Identical immunoglobulin molecules produced by a single clone of plasma cells or cell line.

Programmed cell death Genetically regulated process of self-destruction in certain cells (also known as apoptosis).

PD-1 (programmed death 1) Coinhibitory immune checkpoint receptor involved in suppressing T cell function in peripheral tissues. PD-1 binds two known ligands, PD-L1 and PD-L2.

Progression-free survival The time from random assignment in a clinical trial to disease progression or death from any cause.

Targeted therapy Treatment that is beneficial to a subset of patients (within a given disease) that may be identified by a predictive biomarker determined by a companion diagnostic.

Tyrosine kinase Enzyme that phosphorylates tyrosyl residues on certain proteins. Many receptors (e.g., EGFR) possess this enzymatic activity.

LIST OF ACRONYMS AND ABBREVIATIONS

ADCC Antibody-dependent cellular cytotoxicity
ALK Anaplastic lymphoma kinase
CDC Complement-dependent cytotoxicity
CLL Chronic lymphocytic leukemia
CTLA-4 Cytotoxic T lymphocyte-associated antigen 4
EGFR Epidermal growth factor receptor
EML4 Echinoderm microtubule-associated protein-like 4
Fc Fragment crystallizable

FcγR Fc-gamma receptor
FDA Food and Drug Administration
FISH Fluorescence in situ hybridization
FL Follicular lymphoma
FOXP3 Forkhead box P3
ICOS Inducible T cell costimulator
IHC Immunohistochemistry
KIF5B Kinesin family member 5B
KIT c-Kit proto-oncogene
mAb Monoclonal antibody
MET MET proto-oncogene, receptor tyrosine kinase
NSCLC Non-small cell lung cancer
OS Overall survival
PCD Programmed cell death
PD-1 Programmed death 1
PD-L1 Programmed death ligand-1
PFS Progression-free survival
ROS1 ROS proto-oncogene 1
TFG TRK-fused gene
TKI Tyrosine kinase inhibitor

References

[1] Grillo-Lopez AJ, et al. Overview of the clinical development of rituximab: first monoclonal antibody approved for the treatment of lymphoma. Semin Oncol 1999;26(5 Suppl 14):
66–73.

[2] Minchinton AI, et al. Drug penetration in solid tumours. Nat Rev Cancer 2006;6(8):583–92.

[3] Stashenko P, et al. Characterization of a human B lymphocyte-specific antigen. J Immunol
1980;125(4):1678–85.

[4] Anderson KC, et al. Expression of human B cell-associated antigens on leukemias and
lymphomas: a model of human B cell differentiation. Blood 1984;63(6):1424–33.

[5] Reff ME, et al. Depletion of B cells in vivo by a chimeric mouse human monoclonal
antibody to CD20. Blood 1994;83(2):435–45.

[6] Maloney DG, et al. Phase I clinical trial using escalating single-dose infusion of chimeric anti-CD20 monoclonal antibody (IDEC-C2B8) in patients with recurrent B-cell
lymphoma. Blood 1994;84(8):2457–66.

[7] Maloney DG, et al. IDEC-C2B8 (Rituximab) anti-CD20 monoclonal antibody therapy
in patients with relapsed low-grade non-Hodgkin's lymphoma. Blood 1997;90(6):
2188–95.

[8] McLaughlin P, et al. Rituximab chimeric anti-CD20 monoclonal antibody therapy for
relapsed indolent lymphoma: half of patients respond to a four-dose treatment program. J Clin Oncol 1998;16(8):2825–33.

[9] Dotan E, et al. Impact of rituximab (Rituxan) on the treatment of B cell non-Hodgkin's
lymphoma. P T 2010;35(3):148–57.

[10] Martinelli G, et al. Long-term follow-up of patients with follicular lymphoma receiving single-agent rituximab at two different schedules in trial SAKK 35/98. J Clin Oncol
2010;28(29):4480–4.

[11] Lim SH, et al. Anti-CD20 monoclonal antibodies: historical and future perspectives.
Haematologica 2010;95(1):135–43.

[12] Clynes RA, et al. Inhibitory Fc receptors modulate in vivo cytotoxicity against tumor
targets. Nat Med 2000;6(4):443–6.

[13] Cartron G, et al. Therapeutic activity of humanized anti-CD20 monoclonal antibody and polymorphism in IgG Fc receptor FcgammaRIIIa gene. Blood 2002;99(3):754–8.

[14] Weng WK, et al. Two immunoglobulin G fragment C receptor polymorphisms independently predict response to rituximab in patients with follicular lymphoma. J Clin Oncol 2003;21(21):3940–7.

[15] Farag SS, et al. Fc gamma RIIIa and Fc gamma RIIa polymorphisms do not predict response to rituximab in B-cell chronic lymphocytic leukemia. Blood 2004;103(4):1472–4.

[16] Cartron G, et al. From the bench to the bedside: ways to improve rituximab efficacy. Blood 2004;104(9):2635–42.

[17] Selenko N, et al. Cross-priming of cytotoxic T cells promoted by apoptosis-inducing tumor cell reactive antibodies? J Clin Immunol 2002;22(3):124–30.

[18] Hilchey SP, et al. Rituximab immunotherapy results in the induction of a lymphoma idiotype-specific T-cell response in patients with follicular lymphoma: support for a "vaccinal effect" of rituximab. Blood 2009;113(16):3809–12.

[19] Rezvani AR, et al. Rituximab resistance. Best Pract Res Clin Haematol 2011;24(2):203–16.

[20] Takei K, et al. Analysis of changes in CD20, CD55, and CD59 expression on established rituximab-resistant B-lymphoma cell lines. Leuk Res 2006;30(5):625–31.

[21] Czuczman MS, et al. Acquirement of rituximab resistance in lymphoma cell lines is associated with both global CD20 gene and protein down-regulation regulated at the pretranscriptional and posttranscriptional levels. Clin Cancer Res 2008;14(5):1561–70.

[22] Beers SA, et al. Antigenic modulation limits the efficacy of anti-CD20 antibodies: implications for antibody selection. Blood 2010;115(25):5191–201.

[23] Lim SH, et al. Fc gamma receptor IIb on target B cells promotes rituximab internalization and reduces clinical efficacy. Blood 2011;118(9):2530–40.

[24] Taylor RP, et al. Analyses of CD20 monoclonal antibody-mediated tumor cell killing mechanisms: rational design of dosing strategies. Mol Pharmacol 2014;86(5):485–91.

[25] Kennedy AD, et al. Rituximab infusion promotes rapid complement depletion and acute CD20 loss in chronic lymphocytic leukemia. J Immunol 2004;172(5):3280–8.

[26] Beum PV, et al. The shaving reaction: rituximab/CD20 complexes are removed from mantle cell lymphoma and chronic lymphocytic leukemia cells by THP-1 monocytes. J Immunol 2006;176(4):2600–9.

[27] Li Y, et al. Rituximab-CD20 complexes are shaved from Z138 mantle cell lymphoma cells in intravenous and subcutaneous SCID mouse models. J Immunol 2007;179(6):4263–71.

[28] Williams ME, et al. Thrice-weekly low-dose rituximab decreases CD20 loss via shaving and promotes enhanced targeting in chronic lymphocytic leukemia. J Immunol 2006;177(10):7435–43.

[29] Glennie MJ, et al. Mechanisms of killing by anti-CD20 monoclonal antibodies. Mol Immunol 2007;44(16):3823–37.

[30] Teeling JL, et al. The biological activity of human CD20 monoclonal antibodies is linked to unique epitopes on CD20. J Immunol 2006;177(1):362–71.

[31] Mossner E, et al. Increasing the efficacy of CD20 antibody therapy through the engineering of a new type II anti-CD20 antibody with enhanced direct and immune effector cell-mediated B cell cytotoxicity. Blood 2010;115(22):4393–402.

[32] Goede V, et al. Obinutuzumab plus chlorambucil in patients with CLL and coexisting conditions. N Engl J Med 2014;370(12):1101–10.

[33] Lim SH, et al. Translational medicine in action: anti-CD20 therapy in lymphoma. J Immunol 2014;193(4):1519–24.

[34] Schiller JH, et al. Comparison of four chemotherapy regimens for advanced non-small cell lung cancer. N Engl J Med 2002;346(2):92–8.

[35] Lynch TJ, et al. Activating mutations in the epidermal growth factor receptor underlying responsiveness of non-small cell lung cancer to gefitinib. N Engl J Med 2004;350(21):2129–39.

[36] Paez JG, et al. EGFR mutations in lung cancer: correlation with clinical response to gefitinib therapy. Science 2004;304(5676):1497–500.

[37] Pao W, et al. EGF receptor gene mutations are common in lung cancers from "never smokers" and are associated with sensitivity of tumors to gefitinib and erlotinib. Proc Natl Acad Sci USA 2004;101(36):13306–11.

[38] Soda M, et al. Identification of the transforming EML4–ALK fusion gene in non-small-cell lung cancer. Nature 2007;448(7153):561–6.

[39] Soda M, et al. A mouse model for EML4–ALK-positive lung cancer. Proc Natl Acad Sci USA 2008;105(50):19893–7.

[40] Steuer CE, et al. ALK-positive non-small cell lung cancer: mechanisms of resistance and emerging treatment options. Cancer 2014;120(16):2392–402.

[41] Inamura K, et al. EML4–ALK fusion is linked to histological characteristics in a subset of lung cancers. J Thorac Oncol 2008;3(1):13–7.

[42] Rodig SJ, et al. Unique clinicopathologic features characterize ALK-rearranged lung adenocarcinoma in the Western population. Clin Cancer Res 2009;15(16):5216–23.

[43] Shaw AT, et al. Clinical features and outcome of patients with non-small-cell lung cancer who harbor EML4–ALK. J Clin Oncol 2009;27(26):4247–53.

[44] Takeuchi K, et al. KIF5B–ALK, a novel fusion oncokinase identified by an immuno-histochemistry-based diagnostic system for ALK-positive lung cancer. Clin Cancer Res 2009;15(9):3143–9.

[45] Hernandez L, et al. TRK-fused gene (TFG) is a new partner of ALK in anaplastic large cell lymphoma producing two structurally different TFG-ALK translocations. Blood 1999;94(9):3265–8.

[46] Ou SH, et al. Crizotinib for the treatment of ALK-rearranged non-small cell lung cancer: a success story to usher in the second decade of molecular targeted therapy in oncology. Oncologist 2012;17(11):1351–75.

[47] Christensen JG, et al. Cytoreductive antitumor activity of PF-2341066, a novel inhibitor of anaplastic lymphoma kinase and c-Met, in experimental models of anaplastic large-cell lymphoma. Mol Cancer Ther 2007;6(12 Pt 1):3314–22.

[48] Zou HY, et al. An orally available small-molecule inhibitor of c-Met, PF-2341066, exhibits cytoreductive antitumor efficacy through antiproliferative and antiangiogenic mechanisms. Cancer Res 2007;67(9):4408–17.

[49] Kwak EL, et al. Anaplastic lymphoma kinase inhibition in non-small cell lung cancer. N Engl J Med 2010;363(18):1693–703.

[50] Camidge DR, et al. Activity and safety of crizotinib in patients with ALK-positive non-small-cell lung cancer: updated results from a phase 1 study. Lancet Oncol 2012;13(10):1011–9.

[51] Frampton JE. Crizotinib: a review of its use in the treatment of anaplastic lymphoma kinase-positive, advanced non-small cell lung cancer. Drugs 2013;73(18):2031–51.

[52] Weickhardt AJ, et al. Diagnostic assays for identification of anaplastic lymphoma kinase-positive non-small cell lung cancer. Cancer 2013;119(8):1467–77.

[53] Mosse YP, et al. Safety and activity of crizotinib for paediatric patients with refractory solid tumours or anaplastic large-cell lymphoma: a Children's Oncology Group phase 1 consortium study. Lancet Oncol 2013;14(6):472–80.

[54] Eyre TA, et al. Anaplastic lymphoma kinase-positive anaplastic large cell lymphoma: current and future perspectives in adult and paediatric disease. Eur J Haematol 2014;93(6):455–68.

[55] Vassal G. Biomarker-driven access to crizotinib in ALK-, MET-, or ROS1-positive malignancies in adults and children: Feasibility of the French National Acsé Program. J Clin Oncol 2014;32. 5s (suppl; abstr TPS2647).

[56] Choi YL, et al. EML4–ALK mutations in lung cancer that confer resistance to ALK inhibitors. N Engl J Med 2010;363(18):1734–9.

[57] Friboulet L, et al. The ALK inhibitor ceritinib overcomes crizotinib resistance in non-small cell lung cancer. Cancer Discov 2014;4(6):662–73.

[58] Shaw AT, et al. Ceritinib in ALK-rearranged non-small cell lung cancer. N Engl J Med 2014;370(13):1189–97.

[59] Sakamoto H, et al. CH5424802, a selective ALK inhibitor capable of blocking the resistant gatekeeper mutant. Cancer Cell 2011;19(5):679–90.

[60] Zou HY, et al. PF-06463922 is a potent and selective next-generation ROS1/ALK inhibitor capable of blocking crizotinib-resistant ROS1 mutations. Proc Natl Acad Sci USA 2015;112(11):3493–8.

[61] Pillai RN, et al. Heat shock protein 90 inhibitors in non-small cell lung cancer. Curr Opin Oncol 2014;26(2):159–64.

[62] Sang J, et al. Targeted inhibition of the molecular chaperone Hsp90 overcomes ALK inhibitor resistance in non-small cell lung cancer. Cancer Discov 2013;3(4):430–43.

[63] Krummel MF, et al. CD28 and CTLA-4 have opposing effects on the response of T cells to stimulation. J Exp Med 1995;182(2):459–65.

[64] Rudd CE, et al. CD28 and CTLA-4 coreceptor expression and signal transduction. Immunol Rev 2009;229(1):12–26.

[65] Leach DR, et al. Enhancement of antitumor immunity by CTLA-4 blockade. Science 1996;271(5256):1734–6.

[66] Hodi FS, et al. Improved survival with ipilimumab in patients with metastatic melanoma. N Engl J Med 2010;363(8):711–23.

[67] Pardoll DM. The blockade of immune checkpoints in cancer immunotherapy. Nat Rev Cancer 2012;12(4):252–64.

[68] Patel SP, et al. PD-L1 expression as a predictive biomarker in Cancer immunotherapy. Mol Cancer Ther 2015;14(4):847–56.

[69] Hadrup S, et al. Effector CD4 and CD8 T cells and their role in the tumor microenvironment. Cancer Microenviron 2013;6(2):123–33.

[70] McDermott J, et al. Pembrolizumab: PD-1 inhibition as a therapeutic strategy in cancer. Drugs Today (Barc) 2015;51(1):7–20.

[71] Ai M, et al. Immune checkpoint combinations from mouse to man. Cancer Immunol Immunother 2015. [Epub ahead of print].

[72] Curran MA, et al. PD-1 and CTLA-4 combination blockade expands infiltrating T cells and reduces regulatory T and myeloid cells within B16 melanoma tumors. Proc Natl Acad Sci USA 2010;107(9):4275–80.

[73] Wolchok JD, et al. Nivolumab plus ipilimumab in advanced melanoma. N Engl J Med 2013;369(2):122–33.

[74] Sznol M. Survival, response duration, and activity by BRAF mutation (MT) status of nivolumab (NIVO, anti-PD-1, BMS-936558, ONO-4538) and ipilimumab (IPI) concurrent therapy in advanced melanoma (MEL). J Clin Oncol 2014;32. 5s (suppl; abstr LBA9003^).

[75] Belcaid Z, et al. Focal radiation therapy combined with 4-1BB activation and CTLA-4 blockade yields long-term survival and a protective antigen-specific memory response in a murine glioma model. PLoS One 2014;9(7):e101764.

[76] Curran MA, et al. Combination CTLA-4 blockade and 4-1BB activation enhances tumor rejection by increasing T cell infiltration, proliferation, and cytokine production. PLoS One 2011;6(4):e19499.

[77] Kocak E, et al. Combination therapy with anti-CTL antigen-4 and anti-4-1BB antibodies enhances cancer immunity and reduces autoimmunity. Cancer Res 2006;66(14):7276–84.

[78] Ascierto PA, et al. Clinical experiences with anti-CD137 and anti-PD1 therapeutic antibodies. Semin Oncol 2010;37(5):508–16.

[79] Powles T, et al. MPDL3280A (anti-PD-L1) treatment leads to clinical activity in metastatic bladder cancer. Nature 2014;515(7528):558–62.

[80] Callahan MK. Peripheral and tumor immune correlates in patients with advanced melanoma treated with combination nivolumab (anti-PD-1, BMS-936558, ONO-4538) and ipilimumab. J Clin Oncol 2013;31 (suppl; abstr 3003^).

[81] Ng Tang D, et al. Increased frequency of ICOS+ CD4 T cells as a pharmacodynamic biomarker for anti-CTLA-4 therapy. Cancer Immunol Res 2013;1(4):229–34.

[82] Simeone E, et al. Immunological and biological changes during ipilimumab treatment and their potential correlation with clinical response and survival in patients with advanced melanoma. Cancer Immunol Immunother 2014;63(7):675–83.

[83] Hamid O, et al. A prospective phase II trial exploring the association between tumor microenvironment biomarkers and clinical activity of ipilimumab in advanced melanoma. J Transl Med 2011;9:204.

[84] Ascierto PA, et al. The additional facet of immunoscore: immunoprofiling as a possible predictive tool for cancer treatment. J Transl Med 2013;11:54.

[85] Postow MA. T cell receptor diversity evaluation to predict patient response to Ipilimumab in metastatic melanoma. J Immunother Cancer 2014;2(Suppl 3):O8.

[86] Zamarin D, et al. Immune checkpoint modulation: rational design of combination strategies. Pharmacol Ther 2015;150:23–32.

CHAPTER

7

Translational Approaches in Alzheimer's Disease

Stephen Wood[1], Gabriel Vargas[2]

[1]Neuroscience Discovery Research, Amgen Inc., Thousand Oaks, CA, USA;
[2]Neuroscience Early Development, Amgen Inc., Thousand Oaks, CA, USA

Translational Medicine: Tools and Techniques
http://dx.doi.org/10.1016/B978-0-12-803460-6.00007-6

ALZHEIMER DISEASE AND THE ROLE OF AMYLOID

AD Epidemiology

Alzheimer disease (AD), the most common form of dementia, is a severe and life-threatening neurodegenerative disorder that affects approximately 5 million people in the United States. With the exception of rare familial forms of early-onset AD (FAD), the vast majority (>95%) of people with AD are aged 65 and older. The incidence of AD increases sharply with advancing age. One in nine individuals over the age of 65 has AD. This number increases to 1 in 3 individuals over the age of 85. Approximately two-thirds of Americans with AD are women. In addition, elderly African-Americans and Hispanics are more likely than elderly whites to have AD. The number of individuals afflicted by AD is expected to nearly triple by 2050 as the baby boomer generation in the United States reaches the age of susceptibility. On average, AD patients live 8 years postdiagnosis, with up to 40% of that time requiring full-time care. The socioeconomic impact is expected to reach an excess of US$1 trillion by 2050 if no disease-modifying treatments are found.

AD Pathophysiology

AD was first described by Alois Alzheimer in 1901 [1] and named after him by the eminent psychiatrist Emil Kraepelin. Although there is still some controversy about the cause of this disorder, genetic data have strongly suggested that amyloid deposition is a major determinant [2]. The hallmark features of the disorder are impaired memory along with a loss of executive function leading to problems with activities of daily living.

One of the two neuropathological hallmarks of AD is known as the amyloid plaque. They were characterized during postmortem histological analysis of the first patient [1]. These deposits were initially characterized as starch-like, extracellular deposits in the parenchyma of the AD brain. It would be another three quarters of a century before pioneering work in the field would identify the key component of these plaques as a peptide called amyloid beta or $A\beta$ [3,4]. The $A\beta$ peptide is produced via proteolytic cleavage and release from a larger, precursor protein, the amyloid precursor protein (APP) by the sequential actions of two enzymes, beta-site APP-cleaving enzyme 1 (BACE1) and gamma secretase (Figure 1).

The APP gene lies on chromosome 21, which is of particular interest due to the fact that individuals with trisomy 21 (Down syndrome) have three copies of this chromosome and develop amyloid plaques and Alzheimer dementia typically before the age of 40 [4]. Since $A\beta$ is the main component of plaques seen in AD, it is presumed in turn that excess $A\beta$ is the cause of AD

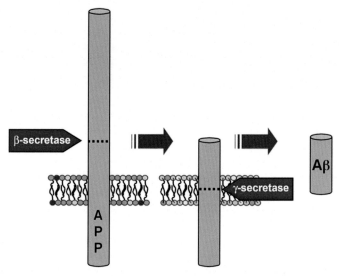

FIGURE 1 The figure denotes the sequential action of beta-secretase and gamma-secretase enzymes acting on the amyloid precursor protein (APP) to generate the Aβ fragment. *Figure by Ipeltan from English Wikipedia (Public domain), via Wikimedia Commons.*

in Down syndrome. Shortly thereafter, it was discovered that mutations in APP led to an early-onset, aggressive form of AD with autosomal dominant inheritance in a Swedish family [5]. Soon after the amyloid hypothesis was formed that proposed simply that amyloid played a key pathological role in AD [2]. Further genetic support for the amyloid hypothesis came shortly after when mutations in presenilin1 (PS1) and presenilin2 (PS2) were found to be linked to autosomal dominant AD [6–8]. PS1 and PS2 are homologous proteins that can each comprise the catalytic activity of gamma secretase [9]. Biochemical analysis of these PS mutations showed that they resulted in an aberrant processing of APP in vivo [10]. In cell culture and in transgenic mouse models, the most common feature of the PS1 and PS2 familial mutations was to increase the amount of Aβ42 produced relative to Aβ40 [11,12]. This is of particular interest in that the longer, Aβ42 fragment, although typically found in lower abundance physiologically, is the form that more readily aggregates in vitro and the predominant form found in amyloid plaques of AD brains [13,14].

Data using synthetic Aβ peptides and cell culture were also supportive. In vitro Aβ readily forms classical beta-pleated sheet aggregates under physiological incubation conditions and these aggregates proved to induce a range of toxic responses to a variety of cell types in culture. Further characterization of this process identified multiple intermediates in the fibrillization process, including soluble oligomeric species. Although the precise molecular identity of these Aβ oligomers remains unsettled, accumulating evidence suggests that soluble forms of Aβ are indeed the proximate effectors of synapse loss and neuronal injury [15]. Finally, transgenic mice overexpressing the human APP containing dominantly inherited mutations demonstrate a time-dependent accumulation of cerebral and cerebrovascular amyloid deposits histologically similar to those observed in AD [16].

Clinical Diagnosis

Traditionally AD was a clinical diagnosis that could not be confirmed until autopsy revealed the hallmark neuropathological features. Recent developments in imaging and molecular biomarkers have led to the updated diagnostic criteria that were recently published [17,18]. The diagnosis of AD is now based on clinical signs and symptoms coupled with the presence of specific biomarkers. This has a potentially profound impact on the way this diagnosis is made in that it is now feasible to have patients diagnosed many years prior to the onset of dementia. However, since there is currently no treatment for AD, the use of biomarkers for diagnostic purposes has remained limited to clinical research settings only. For example, in 2012, the Federal Drug Administration (FDA) approved the use of Amyvid (florbetapir ^{18}F) for positron emission tomographic (PET) imaging to estimate amyloid plaque density in the brains of patients with cognitive impairment who are being evaluated for AD. However, the Centers for Medicare & Medicaid Services denied coverage because of a lack of evidence that the agent could improve health outcomes. The progression of AD is divided into preclinical, prodromal and mild, moderate, and severe AD (Figure 2). The preclinical stage is one where amyloid deposition is present but no cognitive symptoms have emerged, whereas in the prodromal stage, mild but clinically relevant cognitive impairment can be identified. Finally the mild, moderate, and severe AD denotes stages where the patient is already demented with increasing cognitive impairment. However, all these stages must be thought of as a continuum such that once patients have amyloid deposition, they can be considered to have AD.

Current therapies consist mainly of acetylcholinesterase inhibitors and glutamatergic modulators, which provide modest, symptomatic relief. None of these are disease-modifying therapies and thus this area remains one of a large unmet medical need.

THERAPEUTIC APPROACHES

Therapeutic approaches to targeting amyloid in AD have fallen largely into three categories: (1) agents that are designed to prevent the aggregation and formation of amyloid plaques, (2) agents that target the removal

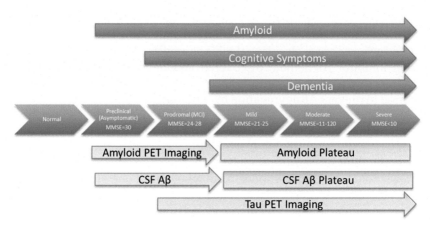

FIGURE 2 The progression of AD is divided into preclinical, prodromal and mild, moderate, and severe Alzheimer disease. The mini-mental status exam (MMSE) is an easy to administer mental exam that can aid in assessing cognitive capacity. Normal score is 30 and anything below that is compromised. The blue arrows denote increasing clinical symptoms. The sandy arrows show increasing signal in the biomarkers and the boxes denote a plateauing of the signal.

TABLE 1 Amyloid-Targeting Therapeutic Approaches

Category	Mechanism of action	Candidates
Plaque formation	Inhibition of amyloid formation	Tramiprosate, ELND005, PBT-1/2
Plaque removal	Active immunization	ACI-24, Affitope AD02, AN-1792, CAD-106
	Passive immunization	AAB-003, aducanumab, BAN2401, bapineuzumab, crenezumab, Gammagard, gantenerumab, LY3002813, MEDI1814, Octagam, ponezumab, SAR228810, solanezumab
Aβ production	γ-secretase inhibition	Avagacestat, semagacestat, Flurizan, CHF 5074, EVP-0962
	β-secretase inhibition	AZD3293, BI 1181181, E2609, JNJ-54861911, LY2886721, MK-8931, RG7129, CNP520

of amyloid plaques and prevent further deposition, and (3) agents that prevent the formation of the precursor of amyloid, the Aβ peptide. Table 1 summarizes amyloid-based therapeutic candidates that have been or are currently being tested in clinical trials. Translational models used preclinically to assess target engagement for amyloid-based therapeutic candidates typically rely on the previously mentioned APP transgenic mouse models as the most reliable and robust phenotype in these mice is the age-dependent deposition of amyloid plaques, which closely resembles the senile amyloid plaques characteristic of AD.

Inhibition of Amyloid Formation

One example of a molecule that inhibits amyloid formation is the aggregation inhibitor Alzhemed, discovered and developed by Neurochem, Inc. The molecule is a variant of the amino acid taurine, which had been reported to prevent Aβ aggregation by inhibiting the interaction of Aβ with glycosaminoglycans. Glycosaminoglycans have been shown to promote Aβ aggregation and plaque stability [19]. In vitro, Alzhemed was shown to preferentially bind soluble Aβ, inhibit Aβ aggregation and fibrillogenesis, and inhibit Aβ neurotoxicity [20]. To assess the effect of Alzhemed in a translational, preclinical model, Neurochem used one of the APP transgenic mouse models, TgCRND8. TgCRND8 mice express a human APP transgene containing the Swedish and Indiana FAD mutations (K670N/M671L and V717F) under the regulation of the Syrian hamster prion promoter to drive neuronal expression. Adult (9 weeks of age) TgCRND8 mice received single, daily subcutaneous injections of 30, 100, or 500 mg/kg Alzhemed for a period of 8–9 weeks. Following treatment, the pharmacodynamic (PD) effect of the compounds was assessed by quantification of the amyloid burden using brain histology and biochemical analysis. For histological analysis, brain slices were stained in 1% Thioflavin S. Thioflavins are dyes used for histology staining and biophysical studies of protein aggregation. It binds to amyloid fibrils but not monomers and gives a distinct increase in fluorescence emission. Another dye that is used to identify amyloid structure is Congo red. Quantitative analysis of amyloid plaque burden included both total number of plaques and percentage of area occupied by plaques. Biochemical analysis of Aβ accumulation was performed by homogenizing brain sections and measuring Aβ content by enzyme-linked immunosorbent assay (ELISA). Soluble and insoluble Aβ fractions were obtained by homogenizing in either Tris buffer or 5M guanidine–HCl, respectively. Systemic treatment with Alzhemed resulted in a significant reduction (roughly 30%) in the brain amyloid plaque load measured histologically and a corresponding approximately 30% reduction in brain Aβ levels measured biochemically [21]. From a pharmacokinetic (PK) standpoint, Alzhemed, dosed peripherally, was shown to cross the blood–brain barrier; however, drug levels in the brain were not quantitatively measured. This is a key component of translational models and will be discussed in this chapter.

Removal of Amyloid Plaques and Prevention of Further Deposition

A second class of amyloid-targeting agents aims to prevent or reverse amyloid formation by harnessing the innate immune system. Active and passive immunization approaches highlight this class. The rationale of the immunotherapy approach is that anti-Aβ antibody binding to amyloid plaques will enable their clearance possibly via activation of microglial phagocytosis. One example is bapineuzumab, codeveloped by Pfizer and Janssen. Bapineuzumab is a humanized form of murine monoclonal antibody 3D6, which targets the N-terminal region of Aβ and binds both fibrillar and soluble Aβ. A large portion of the extensive preclinical work supporting passive anti-Aβ immunotherapy for AD was obtained with the murine parent, 3D6, again, using an APP transgenic mouse model. 3D6 was shown to bind to amyloid plaques in the PDAPP transgenic mouse model and lower plaque burden by 86% compared to an isotype-matched control IgG [22,23]. Immunohistochemical analysis was performed using anti-Aβ antibodies on fixed brain slices.

Inhibition of Aβ Production

Therapeutic candidates that block the production of Aβ fall into the third class of amyloid-targeting agents. As amyloid production is a product of both time and concentration, agents that prevent the formation of Aβ would be predicted to slow or halt the amyloid process. The enzymes responsible for cleaving and releasing Aβ from APP are BACE1 and gamma secretase. Preclinical assessment of these candidates was initially preformed in the APP Tg mice as well. However, as a PD marker of target engagement, acute lowering of Aβ levels in central compartments, i.e., brain and cerebrospinal fluid (CSF) was sufficient and certainly more efficient than chronic dosing studies followed by assessment of plaque burden changes. Additionally, testing Aβ-lowering approaches in transgenic mice may not be optimal in studies of secretase inhibitors since the amount of substrate (APP) and the efficiency of APP cleavage are important factors that can influence Aβ production. Commonly used APP transgenic mice express APP carrying the Swedish mutation, which is cleaved by BACE1 much more efficiently (six- to eightfold) than wild type [24]. Thus, testing BACE1 inhibitors in wild-type mice with physiologically relevant APP expression may provide more meaningful results. BACE inhibitors have been shown to lower endogenous levels of Aβ in brains of mice [25–27]. For these measurements, brain sections are collected following a specified period following dosage of compound and homogenized. Aβ levels are assessed using an ELISA. This requires a pair of antibodies, one recognizing the N-terminal region of Aβ and the other recognizing specifically the C-termini ending at residue 40 (Aβ40) or 42 (Aβ42). Because the rodent (rat and mouse) Aβ sequence differs from the human sequence at three positions within the N-terminal region, rodent-specific N-terminal antibodies are required.

CSF would ultimately become the preferred compartment for assessment of PD coverage for two reasons: (1) PD effects in CSF closely reflect those in brain homogenates and (2) CSF can be obtained from human

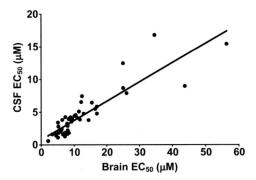

FIGURE 3   In vivo dose–response curves were generated for 48 compounds, and EC_{50} values for brain and CSF Aβ lowering were calculated. A comparison of CSF EC_{50} and brain EC_{50} for these compounds shows a positive, linear relationship between the two compartments ($r2 = 0.778$; $p < 0.0001$; $n = 48$).

subjects in clinical trials, which enables the implementation of a translational biomarker that can be used to determine target engagement even in healthy subjects. Obtaining uncontaminated CSF samples from mice is technically very challenging, however, due to their small size. Therefore, several groups began using larger species such as rats [28], guinea pigs [29], and beagle dogs [30] as preclinical PK-PD models. All are larger than mice making the collection of uncontaminated CSF technically less challenging, however, at the same time requiring more drug substance per study. Data from these preclinical models have shown that CSF and brain Aβ reductions correlate well (Figure 3). In this work, estimates of potency (EC_{50}) for 48 compounds were derived from in vivo dose–response experiments with analysis of Aβ reduction in CSF and brain homogenate. A comparison of the EC_{50} for CSF Aβ reduction with the brain EC_{50} for these compounds displayed a linear relationship with good correlation between the two compartments ($r2 = 0.778$; $p < 0.0001$) [28].

In addition, given a robust enough number of compounds, the rat models provide unique insights into the role that various PK (e.g., liver microsomal turnover, efflux ratio, permeability) and PD (e.g., enzyme and cell potency) parameters play toward enabling Aβ reduction in vivo [28]. Rats have an additional potential advantage as they are typically the rodent species of choice for preclinical toxicity studies, which then allows for the calculation of margins of tox findings to PD effects within a single species rather than across species thus improving the accuracy of this parameter. Guinea pigs have also been used effectively as PK-PD models preclinically [29]. An added feature of the guinea pig model is that their Aβ sequence, unlike mice and rats, is identical to humans eliminating the need for a specialized rat-/mouse-specific ELISA.

The use of nonhuman primates as a preclinical PK-PD model for BACE inhibitors has also been demonstrated [31]. This model, developed in rhesus macaques, allows for direct access to CSF outflow from the cisterna magna via an indwelling catheter [32]. These studies are nonterminal and do not allow for the collection of brain samples. This model does, however, allow for an additional type of analysis of Aβ production and clearance known as stable isotope labeling tandem mass spectrometry (SILT) [33] This method described by Bateman and colleagues utilizes in vivo stable isotope labeling, immunoprecipitation of Aβ from CSF followed by quantitative liquid chromatography–electrospray ionization–tandem mass spectrometry (LC-ESI-tandem MS) to quantify human Aβ protein production and clearance rates. This method has been successfully employed in both nonhuman primates [34] and humans [35].

CLINICAL TRIALS AND BIOMARKERS

Critical Use of Biomarkers for Patient Selection

With the changes to the clinical diagnosis of AD [17,18] and the realization that recent drug failures in the treatment of AD may reflect therapeutic interventions coming too late in the pathological progression, it has been of paramount importance to identify biomarkers that can select patients who will develop the disease. The biomarkers for which the most progress has been made are those that detect amyloid deposition. Amyloid status can be determined either through PET imaging using a variety of ligands such as florbetapir that target amyloid deposition in the brain or through lumbar puncture to measure Aβ levels in the CSF. These two measurements of amyloid have shown remarkable correlation with CSF Aβ levels being inversely correlated with amyloid load as determined by PET imaging [36]. The advent of amyloid PET imaging in the early 2000s with the Pittsburgh compound led to the startling realization that amyloid deposition takes place for more than a decade before symptoms of AD manifest [37]. This has enabled the use of biomarkers to identify subjects who will progress from earlier stages of disease to full-blown dementia. Most clinical trials performed in the past 10 years were done in the dementia stage of the disorder with mild to severe patients. As multiple therapeutic candidates have failed in clinical trials, there has been a push for earlier intervention. This has led to the use of biomarkers to identify subjects at the prodromal stage of the disease. Current efforts by Merck with their BACE inhibitor, MK-8931, and Roche with the gantenerumab program are examples of this approach in which patients are selected based on having amyloid pathology by either PET imaging or CSF sampling for levels of Aβ. One of the realizations from previous AD trials has been that patient populations were not ideally selected and there was no evidence of target engagement in the early development phase that would give use confidence to move to later development. It was later discovered, for example, that in the Phase 3 bapineuzemab APOE4 noncarrier study 36% of participants were found to be amyloid negative (i.e., below the standardized uptake value cut off) at baseline [38]. This is a critical issue in that the therapeutic candidate was targeting amyloid deposition yet more than one-third of the subjects were amyloid negative and therefore would not be expected to benefit from the medication.

Target Engagement Biomarkers

Modern drug development makes evidence of target engagement, one of the key objectives of Phase I clinical trials. A clear detection of target engagement must be demonstrated in order for a compound to be advanced to later development with any degree of conviction. In particular, early development clinical activities aim to answer three general questions: (1) Does the drug get to the target? (2) Does the drug bind to the target? and (3) Does the drug do anything after binding the target? The first question is one of PK, the others refer to target engagement/effect [39]. In terms of the amyloid markers, PET imaging is a perfect example of target binding, whereas measurement of Aβ lowering in the spinal fluid is an example of a PD effect, in this case inhibition of the BACE1 enzyme.

CSF as Target Engagement/PD Biomarker

The development of CSF measurements has allowed for translational approaches in which Aβ levels can be measured in preclinical species as described above and these enable estimation of the dose to be used in humans. Once in humans these translational biomarkers enable a determination of target engagement by measuring Aβ lowering in the spinal fluid in subjects in early development studies. Multiple studies performed by different sponsors have demonstrated that Aβ lowering can be measured in healthy subjects and is not different than what is measured in elderly subjects or those with AD [40–42]. This removes the need to recruit AD patients and thus these studies can be done faster and more cheaply.

One good example of the use of CSF biomarkers was the clinical program for the Lilly BACE1 inhibitor LY2886721 that used preclinical data on CSF Aβ lowering to set their doses for the first-in human study [30]. This was the first example of using translational data for a BACE1 inhibitor. The Lilly BACE1 inhibitor compound LY2886721 demonstrated a robust effect in Aβ lowering in Phase 1 trials in healthy subjects [30]. Measurement of CSF Aβ40 and Aβ42 levels demonstrated a lowering of up to 80% at the 90 mg dose, which was the highest dose tested. The inhibition of Aβ40 was robust and dose dependent and similar to the observations obtained in the preclinical in vivo pharmacology experiments done using beagle dogs [30]. Translational biomarkers aid in development by helping to determine starting doses in humans and more importantly, by having a readout in the clinical trial that the drug is having the desired effect. Without them the development would be guided by reaching a maximum tolerated dose without a clear understanding of whether the drug is having the intended effect. These translational biomarkers are essential to have for evidence-based drug development.

For robust changes to APP processing in man, steady-state measurements of CSF Aβ levels are adequate to quantitate PD effects of secretase inhibitors. For more subtle or transient changes, stable isotope labeling kinetic analysis (SILK) is more appropriate. An example of this is the Lilly gamma secretase inhibitor (LY450139) clinical trial. Attempts to quantitate drug-related Aβ changes were mixed. While some lowering of plasma Aβ was observed, no significant changes were found in the CSF [43]. Using SILK, LY450139 demonstrated a significant decrease in the Aβ production rate in the central nervous system (CNS). Compared to placebo, Aβ production was decreased by 47%, 52%, and 84% at doses of 100, 140, and 280 mg, respectively [44]. Although not as routine and easy as steady-state assessments, this result highlights the increased sensitivity of SILK in detecting changes to CNS APP processing. These results also emphasize the point that plasma Aβ reductions do not necessarily mirror those in the CNS. This relationship needs to be assessed for each therapeutic candidate.

For anti-Aβ antibodies, Aβ42 in the CSF is more useful as a patient selection biomarker and these studies have relied more on imaging data for showing target engagement. Since the antibody therapies are ahead of the BACE inhibitors several of the Phase 3 clinical trials for these therapeutics have published their findings. Unfortunately, clinical trials involving the use of anti-Aβ antibodies have failed to meet their primary end point. Two recent Phase 3 clinical trials with solanezumab, an antibody developed by Lilly that binds to soluble Aβ, failed to meet the primary end points in cognition and function [45]. Posthoc analysis of the pooled data across both studies, however, suggests that perhaps the milder patient populations showed some slowing of cognitive decline when treated with solanezumab. Based on these data, Lilly is continuing development of this candidate.

Similarly the Genentech/Roche antibody crenezumab also showed data in which the primary end point was not met but a post hoc analysis of milder patients suggested a signal in this patient population [46]. While it is generally not a good idea to put too much faith in studies in which a post hoc analysis is necessary to see an effect, the fact that both antibodies showed the same trend is encouraging. In addition, Biogen recently presented data from a prespecified interim analysis of PRIME, the Phase 1b study of aducanumab (BIIB037), which demonstrated encouraging cognitive outcomes in patients with prodromal or mild AD. These data were presented at the 12th International Conference on Alzheimer's and Parkinson's Diseases and Related Neurological

Disorders in Nice, France [47]. It remains to be seen if these data are reproduced in larger clinical trials as the current patient sample size is very small.

For these antibody trials, the use of PET imaging has enabled the determination of target engagement and PD effect. Work done on the development of the Roche anti-Aβ antibody gantenerumab, which binds to Aβ plaques, showed that in patients with mild to moderate AD, infusions of intravenous antibody (60 or 200 mg) for every 4 weeks resulted in a reduction of the levels of brain Aβ amyloid as measured by [^{11}C]PiB PET [48]. These data on the clearing of amyloid by gantenerumab in only 6 months of treatment in a small trial enabled this drug to move forward in development to robust Phase 2/3 trials.

Limitations in the Use of Biomarkers

Recently, however, Roche announced that gantenerumab had failed to meet its primary end point [49]. This is an important point which is that despite demonstrating a PD effect the antibody did not show clinical efficacy. Thus having a PD effect, while essential, is not sufficient. This argues for the need to use biomarkers in early development as an indicator of the minimum drug level required to have an effect. The challenge is to determine how much higher exposure from this minimum is needed to obtain clinical efficacy. The relationship between PD effect and clinical end point is of course unknown until drugs are fully developed and one can demonstrate how these parameters are related. Thus in the development of novel mechanism of action drugs, while it is essential to use biomarkers to show a PD effect, it is recommended to go as high as tolerated unless there are data present which suggest that the relationship between the target engagement and PD biomarkers to the clinical end point is apparent. This is not usually the case and many programs do not explore a high enough dose.

CONCLUDING COMMENTS AND FUTURE DIRECTIONS

The field of AD therapy has been working to develop disease-modifying drugs for decades now without much success. In order to remedy this situation an approach that integrates biomarkers is essential. The development of translational biomarkers has enabled progress in the three major areas discussed in this chapter: (1) diagnosis of AD, (2) patient selection for clinical trials, and (3) target engagement and PD assessments.

Biomarkers of amyloid load have permitted the identification of AD pathology years and even decades before the onset of dementia. These biomarkers in turn have aided in the development of new criteria for the diagnosis and patient selection for clinical trials targeting amyloid. Without these biomarkers the field would still be intervening once dementia sets in, which is most likely too late in the disease process to have much of an impact. Earlier intervention is expected to have a positive effect on outcomes. A particular challenge of moving into earlier stages of the disease is that capturing clinical benefit becomes even more difficult as the extent of the clinical pathology diminishes the earlier one attempts to intervene.

The use of these biomarkers for clinical trial selection is very important as they enable recruitment into clinical trials only for those patients with the right pathology and exclude those who would not benefit from the therapy. This has the potential to increase the power of the study and improves the risk benefit for those patients who are enrolled. This in short is the definition of personalized health care: right patient, at the right time, with the right drug.

Once in clinical trials, target engagement must be demonstrated for the therapeutic compound. Much work done in the preclinical space in rodents and nonhuman primates show that CSF Aβ is a useful translational biomarker that can be used for PD effect/target engagement assessment.

Moving forward the ideal approach toward AD drug development would be to use surrogate markers of drug effect. If one can show that biomarkers correlate with clinical improvement, this would enable future trials that use these biomarkers as their primary end point. This would be a huge advancement for the field. The FDA recently published draft guidance indicating in principle the viability of this approach [49].

Much work is currently being undertaken in the development of tracers for tau pathology, which has been shown previously to correlate with cognitive deficits to a greater extent than any measures of amyloid load. In the next several years we will see this development in tau imaging emerge much like amyloid imaging did about 10 years ago. This development along with many others in the field of translational biomarkers will have a positive impact on the development of therapeutics aimed at having a disease-modifying effect on this severe neurodegenerative disorder.

Glossary

Aβ Fragment of APP and main component of amyloid plaques, a key pathological hallmark observed in the brains of individuals with Alzheimer disease.

Amyloid Insoluble protein aggregates exhibiting a beta sheet structure.

Amyloid precursor protein (APP) An integral membrane protein expressed in many tissues and concentrated in the synapses of neurons.

Beta-site APP-cleaving enzyme (BACE) Aspartyl protease responsible for the rate-limiting step in the production of Aβ from APP.

Cerebrospinal fluid (CSF) A clear, colorless fluid found in the central nervous system. It is produced in the choroid plexus and circulates throughout the brain and spinal cord.

Gamma secretase A multisubunit, intramembrane protease complex that cleaves single-pass transmembrane proteins, including APP, at residues within the transmembrane domain.

Mini-mental state examination (MMSE) A 30-point questionnaire that is used in clinical and research settings to measure cognitive impairment. It is commonly used to screen for dementia.

Positron emission tomography (PET) A functional imaging technique that produces a three-dimensional image in the body. The system detects pairs of gamma rays emitted indirectly by a positron-emitting radionuclide (tracer), which is introduced into the body on a biologically active molecule. Three-dimensional images of tracer concentration within the body are then constructed by computer analysis.

LIST OF ACRONYMS AND ABBREVIATIONS

APP Amyloid precursor protein
Aβ Amyloid beta peptide
BACE Beta-site APP-cleaving enzyme
CNS Central nervous system
CSF Cerebrospinal fluid
ELISA Enzyme-linked immunosorbent assay
FAD Familial Alzheimer disease
FDA Federal Drug Administration
MMSE Mini-mental state examination
NIH National Institutes of Health
PD Pharmacodynamic
PET Positron emission tomography
PK Pharmacokinetic
SILK Stable isotope labeled kinetic analysis
SILT Stable isotope labeling tandem mass spectrometry

References

[1] Alzheimer A. About a peculiar disease of the cerebral cortex. Cent fur Nervenheilkd Psychiatr 1907;30:177–9.

[2] Hardy JA, Higgins GA. Alzheimer's disease: the amyloid cascade hypothesis. Science 1992;256(5054):184–5.

[3] Allsop D, Landon M, Kidd M. The isolation and amino acid composition of senile plaque core protein. Brain Res 1983;259(2):348–52.

[4] Glenner GG, Wong CW. Alzheimer's disease and Down's syndrome: sharing of a unique cerebrovascular amyloid fibril protein. Biochem Biophys Res Commun 1984;122(3):1131–5.

[5] Goate A, Chartier-Harlin MC, Mullan M, Brown J, Crawford F, Fidani L, et al. Segregation of a missense mutation in the amyloid precursor protein gene with familial Alzheimer's disease. Nature 1991;349(6311):704–6.

[6] Levy-Lahad E, Wasco W, Poorkaj P, Romano DM, Oshima J, Pettingell WH, et al. Candidate gene for the chromosome 1 familial Alzheimer's disease locus. Science 1995;269(5226):973–7.

[7] Rogaev EI, Sherrington R, Rogaeva EA, Levesque G, Ikeda M, Liang Y, et al. Familial Alzheimer's disease in kindreds with missense mutations in a gene on chromosome 1 related to the Alzheimer's disease type 3 gene. Nature 1995;376(6543):775–8.

[8] Sherrington R, Rogaev EI, Liang Y, Rogaeva EA, Levesque G, Ikeda M, et al. Cloning of a gene bearing missense mutations in early-onset familial Alzheimer's disease. Nature 1995;375(6534):754–60.

[9] Wolfe MS, Xia W, Ostaszewski BL, Diehl TS, Kimberly WT, Selkoe DJ. Two transmembrane aspartates in presenilin-1 required for presenilin endoproteolysis and gamma-secretase activity. Nature 1999;398(6727):513–7.

[10] Scheuner D, Eckman C, Jensen M, Song X, Citron M, Suzuki N, et al. Secreted amyloid beta-protein similar to that in the senile plaques of Alzheimer's disease is increased in vivo by the presenilin 1 and 2 and APP mutations linked to familial Alzheimer's disease. Nat Med 1996;2(8):864–70.

[11] Borchelt DR, Thinakaran G, Eckman CB, Lee MK, Davenport F, Ratovitsky T, et al. Familial Alzheimer's disease-linked presenilin 1 variants elevate Abeta1-42/1-40 ratio in vitro and in vivo. Neuron 1996;17(5):1005–13.

[12] Citron M, Westaway D, Xia W, Carlson G, Diehl T, Levesque G, et al. Mutant presenilins of Alzheimer's disease increase production of 42-residue amyloid beta-protein in both transfected cells and transgenic mice. Nat Med 1997;3(1):67–72.

[13] Jarrett JT, Berger EP, Lansbury Jr PT. The C-terminus of the beta protein is critical in amyloidogenesis. Ann N Y Acad Sci 1993;695:144–8.

[14] Welander H, Franberg J, Graff C, Sundstrom E, Winblad B, Tjernberg LO. Abeta43 is more frequent than Abeta40 in amyloid plaque cores from Alzheimer disease brains. J Neurochem 2009;110(2):697–706.

[15] Walsh DM, Selkoe DJ. A beta oligomers - a decade of discovery. J Neurochem 2007; 101(5):1172–84.

[16] Balducci C, Forloni G. APP transgenic mice: their use and limitations. Neuromol Med 2011;13(2):117–37.

[17] Dubois B, Feldman HH, Jacova C, Cummings JL, Dekosky ST, Barberger-Gateau P, et al. Revising the definition of Alzheimer's disease: a new lexicon. Lancet Neurol 2010;9(11):1118–27.

[18] Dubois B, Feldman HH, Jacova C, Hampel H, Molinuevo JL, Blennow K, et al. Advancing research diagnostic criteria for Alzheimer's disease: the IWG-2 criteria. Lancet Neurol 2014;13(6):614–29.

[19] Gupta-Bansal R, Frederickson RC, Brunden KR. Proteoglycan-mediated inhibition of A beta proteolysis. A potential cause of senile plaque accumulation. J Biol Chem 1995;270(31):18666–71.

[20] Gervais F, Chalifour R, Garceau D, Kong X, Laurin J, McLaughlin R, et al. Glycosaminoglycan mimetics: a therapeutic approach to cerebral amyloid angiopathy. Amyloid 2001;8(Suppl. 1):28–35.

[21] Gervais F, Paquette J, Morissette C, Krzywkowski P, Yu M, Azzi M, et al. Targeting soluble Abeta peptide with Tramiprosate for the treatment of brain amyloidosis. Neurobiol Aging 2007;28(4):537–47.

[22] Bard F, Barbour R, Cannon C, Carretto R, Fox M, Games D, et al. Epitope and isotype specificities of antibodies to beta -amyloid peptide for protection against Alzheimer's disease-like neuropathology. Proc Natl Acad Sci 2004;100(4):2023–8.

[23] Bard F, Cannon C, Barbour R, Burke RL, Games D, Grajeda H, et al. Peripherally administered antibodies against amyloid beta-peptide enter the central nervous system and reduce pathology in a mouse model of Alzheimer disease. Nat Med 2000;6(8):916–9.

[24] Citron M, Oltersdorf T, Haass C, McConlogue L, Hung AY, Seubert P, et al. Mutation of the beta-amyloid precursor protein in familial Alzheimer's disease increases beta-protein production. Nature 1992;360(6405):672–4.

[25] Jeppsson F, Eketjall S, Janson J, Karlstrom S, Gustavsson S, Olsson LL, et al. Discovery of AZD3839, a potent and selective BACE1 inhibitor clinical candidate for the treatment of Alzheimer disease. J Biol Chem 2012;287(49):41245–57.

[26] Nishitomi K, Sakaguchi G, Horikoshi Y, Gray AJ, Maeda M, Hirata-Fukae C, et al. BACE1 inhibition reduces endogenous Abeta and alters APP processing in wild-type mice. J Neurochem 2006;99(6):1555–63.

[27] Swahn BM, Kolmodin K, Karlstrom S, von Berg S, Soderman P, Holenz J, et al. Design and synthesis of beta-site amyloid precursor protein cleaving enzyme (BACE1) inhibitors with in vivo brain reduction of beta-amyloid peptides. J Med Chem 2012; 55(21):9346–61.

[28] Wood S, Wen PH, Zhang J, Zhu L, Luo Y, Babu-Khan S, et al. Establishing the relationship between in vitro potency, pharmacokinetic, and pharmacodynamic parameters in a series of orally available, hydroxyethylamine-derived beta-secretase inhibitors. J Pharmacol Exp Ther 2012;343(2):460–7.

[29] Janson J, Eketjall S, Tunblad K, Jeppsson F, Von Berg S, Niva C, et al. Population PKPD modeling of BACE1 inhibitor-induced reduction in Abeta levels in vivo and correlation to in vitro potency in primary cortical neurons from mouse and guinea pig. Pharm Res 2014;31(3):670–83.

[30] May PC, Willis BA, Lowe SL, Dean RA, Monk SA, Cocke PJ, et al. The potent BACE1 inhibitor LY2886721 elicits robust central Abeta pharmacodynamic responses in mice, dogs, and humans. J Neurosci 2015;35(3):1199–210.

[31] Sankaranarayanan S, Holahan MA, Colussi D, Crouthamel MC, Devanarayan V, Ellis J, et al. First demonstration of cerebrospinal fluid and plasma A beta lowering with oral administration of a beta-site amyloid precursor protein-cleaving enzyme 1 inhibitor in nonhuman primates. J Pharmacol Exp Ther 2009;328(1):131–40.

[32] Gilberto DB, Zeoli AH, Szczerba PJ, Gehret JR, Holahan MA, Sitko GR, et al. An alternative method of chronic cerebrospinal fluid collection via the cisterna magna in conscious rhesus monkeys. Contemp Top Lab Anim Sci 2003;42(4):53–9.

[33] Bateman RJ, Munsell LY, Chen X, Holtzman DM, Yarasheski KE. Stable isotope labeling tandem mass spectrometry (SILT) to quantify protein production and clearance rates. J Am Soc Mass Spectrom 2007;18(6):997–1006.

[34] Cook JJ, Wildsmith KR, Gilberto DB, Holahan MA, Kinney GG, Mathers PD, et al. Acute gamma-secretase inhibition of nonhuman primate CNS shifts amyloid precursor protein (APP) metabolism from amyloid-beta production to alternative APP fragments without amyloid-beta rebound. J Neurosci 2010;30(19):6743–50.

[35] Bateman RJ, Munsell LY, Morris JC, Swarm R, Yarasheski KE, Holtzman DM. Human amyloid-beta synthesis and clearance rates as measured in cerebrospinal fluid in vivo. Nat Med 2006;12(7):856–61.

[36] Fagan AM, Mintun MA, Mach RH, Lee SY, Dence CS, Shah AR, et al. Inverse relation between in vivo amyloid imaging load and cerebrospinal fluid Abeta42 in humans. Ann Neurol 2006;59(3):512–9.

[37] Klunk WE, Engler H, Nordberg A, Wang Y, Blomqvist G, Holt DP, et al. Imaging brain amyloid in Alzheimer's disease with Pittsburgh Compound-B. Ann Neurol 2004;55(3):306–19.

[38] Salloway S, Sperling R, Brashear HR. Phase 3 trials of solanezumab and bapineuzumab for Alzheimer's disease. N Engl J Med 2014;370(15):1460.

[39] Morgan P, Van Der Graaf PH, Arrowsmith J, Feltner DE, Drummond KS, Wegner CD, et al. Can the flow of medicines be improved? Fundamental pharmacokinetic and pharmacological principles toward improving phase II survival. Drug Discov Today 2012;17(9–10):419–24.

[40] The novel BACE inhibitor MK-8931 dramatically lowers CSF Abeta peptide in patients with mild to moderate Alzheimer's disease. In: Forman M, Tseng J, Dockendorf M, Canales C, Apter J, Backonja M, editors. The 11th international conference on Alzheimer's and Parkinson's diseases. 2013. Florence, Italy.

[41] The novel BACE inhibitor MK-8931 dramatically lowers CSF Aβ peptides in healthy subjects: results from a rising single dose study. In: Forman M, Tseng J, Palcza J, Leempoels J, Ramael S, Krishna G, editors. 64th American Academy of Neurology annual meeting. 2012. New Orleans, LA.

[42] First-in-human study of E2609, a novel BACE1 inhibitor, demonstrates prolonged reductions in plasma beta-amyloid levels after single dosing. In: Lai R, Albala B, Kaplow JM, Aluri J, Yen M, Satlin A, editors. The 11th international conference on Alzheimer's & Parkinson's diseases. 2013. Florence, Italy.

[43] Siemers ER, Quinn JF, Kaye J, Farlow MR, Porsteinsson A, Tariot P, et al. Effects of a gamma-secretase inhibitor in a randomized study of patients with Alzheimer disease. Neurology 2006;66(4):602–4.

[44] Bateman RJ, Siemers ER, Mawuenyega KG, Wen G, Browning KR, Sigurdson WC, et al. A gamma-secretase inhibitor decreases amyloid-beta production in the central nervous system. Ann Neurol 2009;66(1):48–54.

[45] Doody RS, Thomas RG, Farlow M, Iwatsubo T, Vellas B, Joffe S, et al. Phase 3 trials of solanezumab for mild-to-moderate Alzheimer's disease. N Engl J Med 2014;370(4):311–21.

[46] Investor Update [press release]. July 16, 2014.

[47] Randomized, double-blind, phase 1B study of BIIB037, an anti-amyloid beta monoclonal antibody, in patients with prodromal or mild Alzheimer's disease. In: Sevigny J, Chiao P, Williams L, Miao X, O'Gorman J, editors. The 12th International Conference on Azheimer's and Parkinson's Diseases. 2015. Nice, France.

[48] Ostrowitzki S, Deptula D, Thurfjell L, Barkhof F, Bohrmann B, Brooks DJ, et al. Mechanism of amyloid removal in patients with Alzheimer disease treated with gantenerumab. Arch Neurol 2012;69(2):198–207.

[49] FDA Draft Guidance for Industry; Alzheimer's Disease: Developing Drugs for the Treatment of Early Stage Disease; 2014.

Index